# ECONOMICS AND HOME PRODUCTION

For my wife, Juat Mei,
and children, Josh and Jo-Ann.

# Economics and Home Production

Theory and measurement

EUSTON QUAH
*Department of Economics and Statistics*
*National University of Singapore*

# Avebury

Aldershot · Brookfield USA · Hong Kong · Singapore · Sydney

Published by
Avebury
Ashgate Publishing Limited
Gower House
Croft Road
Aldershot
Hants GU11 3HR
England

Ashgate Publishing Company
Old Post Road
Brookfield
Vermont 05036
USA


**British Library Cataloguing in Publication Data**

Quah, Euston
    Economics and Home Production:
    Theory and Measurement
    I. Title
    338.6

ISBN 1 85628 457 3




Printed and Bound in Great Britain by
Athenaeum Press Ltd, Newcastle upon Tyne.

# Contents

v

vi

# List of illustrations

# List of tables

# Acknowledgements

The list of people with whom I have worked closely in writing this study and to whom I owe much is a long one. Foremost on this list are Emeritus Professor Lim Chong Yah, formerly Head, Department of Economics and Statistics, and Dr Koh Ai Tee, both of the National University of Singapore. Professor Lim has provided much support, encouragement and advice throughout. I am also grateful to him for reducing my teaching load over the course of my writing this study, and thus providing the appropriate environment in which I could do my thinking.

My intellectual debt to Dr Koh Ai Tee is immeasurable. I owe her a special debt in tactfully increasing and improving the sense of many of my sentences. I am also grateful for her patience and willingness to spend countless hours with me discussing problems, pointing out omissions and correcting many errors of earlier drafts of the present study.

I have also benefitted greatly from earlier discussions with my colleagues: Dr Hoon Hian Teck, Associate Professor Bhanoji Rao, Dr Tan Khye Chong (now of Nanyang Technological University, Singapore) and Dr Tse Yiu Kuen, the latter two provided much helpful econometric sense. I am also indebted to Professors Luisella Goldschmidt-Clermont of the International Labour Organisation at Geneva, Switzerland, Keith Bryant of Cornell University, Catherine Zick of the University of Utah, Jack Knetsch of Simon Fraser University, Malcolm Rutherford of the University of Victoria, Canada and Ezra Mishan formerly of the London School of Economics for some stimulating productive exchanges based on mutual research interests.

I am also grateful to Professor Herbert Grubel of Simon Fraser University for his perceptive comments and discussions with me on this research topic during his visit to Singapore as Distinguished Visitor of Banking and International Finance at the Institute of Southeast Asian

xiv

Studies.
Preliminary theoretical work of this study was presented in the Department's Economic Theory and Modelling Workshop, for which I was grateful to member-colleagues especially Associate Professor Basant Kapur and Dr David Evans (now of Australian National University) for helpful comments.

Special debts are also owed to my colleagues Drs Lin Ting Kwong and G Shantakumar for greatly aiding the drafting of the survey questionnaire used in this study. Many of the ambiguities contained in earlier drafts of the survey questions were removed and the sense increased considerably. Also, I would like to thank Mr Tung Siew Hoong for his assistance in computer work, and to Mr Isaac Koh for helpful preparation in the organisation of this manuscript.

I am also grateful to Mrs Leow Bee Geok, Director of Research and Principal Statistician at the Ministry of Labour and Department of Statistics respectively who provided invaluable comments on survey design, and to Miss Wong Mei Kwong and Major Lim, Assistant Directors at the Ministry of Labour (Work Permit Division) and Miss Long Chu, Senior Statistician of the National Accounts Division, Department of Statistics for helpful discussion, information and cooperation. And, to Mr Jiuan of the Embassy of Finland in Singapore for supplying information on work done in Finland on related aspects of household production.

I also wish to acknowledge the kindness of the Editors of the American Journal of Economics and Sociology, Applied Economics, Asian Economic Journal and the Osgoode Hall Law Journal for permission to make use of the ideas and exposition in my papers which appeared in these journals.

Finally, I wish to acknowledge the financial assistance and support in the form of a Research Grant from the National University of Singapore. Without this support, the household survey would have been extremely difficult to undertake. And, not forgetting the competent and efficient typing and organisational work on the survey by Mrs Elsie Koo, Miss Ham Oi Mun and Ms Kamariah Sumshuddin and to my secretary, Mrs June Lee for typing the final form of the manuscript.

All of these people mentioned above deserve much credit for whatever merit this book contains. The remaining errors and misinterpretations are solely mine.

**Euston Quah**, PhD
National University of Singapore,
1993

# Preface

This book deals with recent developments in economic theory pertaining to the study of household production and valuation. It examines the major theoretical and methodological issues involved in measuring and valuing household production (or household work) such as definition, quantification, joint-production, and valuation. This is an important area of research in household economics, and this book goes well beyond earlier efforts in dealing effectively with serious and controversial measurement issues.

The book also provides a comprehensive and up-to-date survey of major empirical studies done on measuring household production in the United States, Canada, Europe and Asia. The methods used, results and limitations are reported. The studies are chronologically dated and listed by country origins.

A key element in the book's contents is the author's suggestion of a new way to measure and evaluate household production for social accounting purposes. This takes the form of constructing an efficiency-adjusted index for replacement cost measures, and for the first time, integrating both the well-known replacement cost and opportunity cost approaches.

This book should be very useful to researchers in household economics including family and home economists, sociologists, social workers, national income accountants, and those teaching in the area of microeconomics, women studies, social accounting, as well as students and other related researchers.

# Summary

National income accounts have consistently and to a large extent ignored non-market economic activities. This results in a great deal of misinformation and deficiencies in the final set of estimates. The study investigates the meaning and viability of deriving dollar estimates for the major type of economically productive but non-market activity namely that of household production.

While certainly an old problem revisited, it was only until recently that renewed interest in the accounting of such non marketed household production activities occurred resulting from several progresses made in economic theory (the 'New Home Economics'), data availability and improvements in data collection methods, and the increased desirability to reflect social welfare concerns as well as changes in the structural economy (from non-market to market) in the product accounts.

Despite some of these advances, there have remained some very serious conceptual and technical difficulties in measuring and valuing household production. Thus, difficulties exist with household production definition, quantification and the measurement of joint-production activities. Further, there appears to be much confusion over what is to be measured and hence valued.

Consequently, the work first explores some of the theoretical and methodological issues behind household production research and suggests ways of resolving them. This departs from past studies which were either very empirically or mathematically based and which paid very little if not no attention to these problems.

The study also suggests a new way of evaluating household production for social accounting purposes. The method suggested is shown to be consistent with social accounting practises. This is done through a mathematical model which formally shows that for social accounting purposes, it is the efficiency-adjusted replacement cost method that appears appropriate.

This new method of evaluating household production is then applied to Singapore household data gathered from a survey involving 684 households. The results are then compared to past studies using the conventional valuation approaches. The empirical estimates generated on household production in Singapore further provide a basis for future research since to date, there has been no work done on the non-market household economy of Singapore. It is thus the first comprehensive and systematic study on household production in Singapore.

Such questions as: What is the order of magnitude of the value and quantity of household production in Singapore? What patterns of time-use exist in Singapore? Who are involved in household production and what is the nature of the goods and services produced? To what extent has capital substituted for labour within the home? How has changing home production patterns affected labour force participation rates for women? These and many other macro-micro questions are discussed based on the empirical results of the household survey.

# 1 Introduction

## 1.1 Objectives and scope of this study

### 1.1.1 Objectives

Much of what goes on within households are unrecorded. One important productive activity is that of household work. The amount and value of time and effort used to provide the day-to-day services of cooking, cleaning, shopping, child caring and the myriad other chores that are demanded by households are clearly not insignificant.[1] It represents a very substantial portion of the total productive time and effort available to members of a household and to a society (especially less developed countries given the lower degree of division of labour, monetization and greater intensity of labour inputs[2]) and, an accounting of its quantity and

---

[1] Some studies have reported the amount of time spent in household production to be more than 47 hours a week for employed-wife households to 84 hours a week for unemployed-wife households (Hawrylyshyn, 1978) for Canada; close to 50 hours a week for employed-wife households and 65 hours a week for unemployed-wife households (Peskin, 1976; Walker and Gauger, 1973) for the United States; and between 42 to 66 hours a week per household (Evenson et al., 1980) for the Philippines. Thus, the range of time devoted to household production is quite similar if not more than the time devoted to paid market work. On the value of time devoted to household production, some studies have indicated that this was more than 44 percent of the U.S. GNP in 1976 (Peskin, 1976) and more than 33 percent of the GNP of Canada in 1971 (Hawrylyshyn, 1978). For more empirical estimates on the quantity and value of household production of various studies, see Chapter 4.

[2] For these less developed countries, the definition of household production is much wider and involves subsistence activities like crop farming, collecting firewood, (continued...)

1

economic value are issues of increasing practical importance.

The concerns with the valuation of household production include a social accounting of this significant but uncounted economic production, measuring or weighting the contributions of individual members of a household to their common welfare, and assessing damages in cases involving loss of household production services due to wrongful injury or death. This study addresses the first concern.[3] While certainly on old problem revisited -- economists have been concerned for many years with the delimitation of what is and what is not to be included in national income accounting -- it was only until recently that renewed interest in the accounting of such non-marketed activities occurred resulting from several progresses made in economic theory (especially the New Home Economics[4]), data availability and improvements in data collection methods (the Cornell and Michigan time-use studies[5]), and the increased desirability to reflect social welfare concerns as well as changes in the

---

[2] (...continued)
fishing and some handicrafts. The present study however excludes such activities and adopts a narrower definition of household production. For a fuller discussion on definitional issues, see Chapters 2 and 3.

[3] On the valuation of household production for family welfare purposes, see my paper, 'Valuing Family Household Production: A Contingent Evaluation Approach' in *Applied Economics*, 1987. And, on the valuation of household production for personal injury or wrongful death litigation, see my paper, 'Exact Compensation in Law and Economics: The Theory of Welfare Loss in Household Production' in *Osgoode Hall Law Journal*, 1987.

[4] While traditional micro theory views the household as a consuming unit in society, maximising a utility function that is defined in terms of goods and services bought from the market, modern microeconomics however, regards the household as a multiperson production unit engaged in the production of utility-yielding household goods and services, using market purchased goods and the time of family members as factor inputs. See Becker, 1965; Michael and Becker, 1973; Gronau, 1973; and Ghez and Becker, 1975.

[5] The Cornell University team, of which Walker is the best-known representative, developed a series of time-budget studies on household work for which the data collected was then used by several researchers analyzing household production and behaviour. See Walker, 1969; Walker and Gauger, 1973; and Walker and Woods, 1976. The other well-known collections on time-use is led by Juster and his team of researchers at the Survey Research Centre of the University of Michigan. See Juster et al., 1978. More of this and other time-budget studies in Chapter 3, Section 3.2.

structural economy (from non-market to market) in the conventional national income and product accounts.[6] The latter would aid in better understanding of intertemporal economic performance.

There have however remained some very serious conceptual and technical difficulties in measuring and valuing household production. Thus for example, difficulties exist with the definition and hence quantification of household production, particularly on the identification of production and consumption activities in the home and the measurement of joint-production or simultaneous activities. Another concerns the large heterogeneity in household work and in the quality and efficiency between own home services and market provided home services (eg. domestic help) that make imputations of the value of household production by market equivalents difficult. Difficulties also exist in imputations by opportunity cost of time spent in home production due to the existence of conventional rigidities in time schedules in the employment market.[7] In addition, much confusion exists over what is to be measured and for what purpose. At this point, it appears that, if the objective of valuation is for social accounting, then to be consistent with marketed outputs measurement, the use of market prices or marginal valuation would seem more appropriate. However, for many welfare and compensation issues, the appropriate measurement is one of economic surplus or net benefit measures. The appropriate measures or values to be adopted would thus be sensitive to such varied objectives.[8]

---

[6] Since the study by Nordhaus and Tobin, 1972 which questioned the limitations and of the use of GNP as an indicator of welfare, much attention has been drawn to modifications of the national income and product accounts, see Chapter 4.

[7] Significant problems in imputation using the opportunity cost method also exist for those household members who lacked labour force participation and those who have been out of the labour market for a considerable period of time. In this case, by equating the value of their time spent in home production to their opportunity income may lead to an assertion of near zero dollars. Clearly this cannot be the case. More on this and other problems using the opportunity cost method in Chapter 3.

[8] Thus, for example one finds in the literature that the method of opportunity cost is used to measure the value of household production for cases involving losses of welfare from household production due to wrongful injury or death (Komesar, 1974; Pottick, 1978). But since opportunity cost is not a welfare measure, to assert it as one is theoretically incorrect. Similarly, in litigation cases involving the loss
(continued...)

This study addresses some very pertinent conceptual and methodological issues behind household production research. The study first explores some of the theoretical problems involved in valuing non-market household production and seeks to determine what is in some sense the appropriate valuation for the different purposes of measurement. Many of the quantitative results of some of the previous studies on household production continue to be widely regarded as curiosities and poorly substantiated assertions, and in some cases, obvious flaws exist. As mentioned earlier, no doubt, one important reason for this is that some of the most basic conceptual, theoretical and methodological problems are still unresolved. Further, some of the models developed for household production have been too abstract -- the requirement of unduly restrictive assumptions and the difficulty in trying to operationalize some of the critical variables used in those models -- with the result that not much use (or sense!) can be made of them on the practical front. Indeed, some of the restrictions placed on some of the models are even contrary to casual observations. The work analyzes each of these problems and suggests ways to resolve them.

Second, we attempt to provide a general framework for the valuation of non-market household production consistent with the way market production of goods and services are valued. Here, the study defines household production as comprising of three types: market replaceable household production (MHP), near-market replaceable household production (NMHP) and non-replaceable household production (NHP). MHP consists of those unpaid home activities performed for and by household members resulting in home goods and services which have market equivalents. Examples of such MHP activities are cooking, cleaning and laundry work. Clearly, these activities are also found in the market in the form of domestic help services and therefore identifiable and replaceable. NMHP are those unpaid home activities which are

---

[8](...continued)
of household production services and hence, welfare, the use of the replacement cost method is also theoretically incorrect (in Franco v Woolfe, a wrongful death case, a professor of economics testified that the value of household production is that of an imputed replacement cost estimate for the country's GNP; since the objective of measurement here is for losses in household welfare, the use of an imputed value of household production for GNP purposes is again theoretically inappropriate). Also, see my paper, 'Persistent Problems in Measuring Household Production' in *The American Journal of Economics and Sociology*, 1986.

4

performed in addition to MHP and consists of those activities which are normally not associated with nor replaceable by hiring domestic help but yet can conceivably be done by employing other appropriate market substitutes. Examples of NMHP include, tutoring a child, organising and supervision of household tasks related to family requirements and counselling or giving advice to family members. Finally, NHP consists of those home produced goods and services that have no close market substitutes and cannot be replaced by purchasing market goods and services. Examples of NHP are the love, care and companionship provided by family members for each other.[9]

The value of household production in an economy is derived by summing up the values of MHP and NMHP for all households. The value of NHP need not be taken into account since for the most part they involve much broader non-economic issues (defined narrowly) and ones which clearly lie outside the perimeter of possible imputations of non-market goods and services consistent with the social income accounting framework.

In valuing MHP, a general model of MHP is presented showing a way of conceptualising the household's behaviour. The model differs from previous formulations of household production in that it incorporates the notion of domestic help drawn from the market substituting for some of the self-performed non-market household production, an area commonly neglected in existing formulations. Another distinguishing feature of the general model is that the household's utility depends directly on household production services, rather than just on their contributions to consumption. The model also indicates some of the other limitations of the conventional approaches -- the requirement of restrictive assumptions concerning the utility of market and homework -- and establishes a theoretical basis of valuation of market replaceable household production from its marginal conditions so that imputations for social accounting purposes can be derived. The results from the model show that while using the market replacement wage rate of domestic help may serve as a useful approximation of the value of market replaceable household production, a more accurate imputation would depend on whether the household is as efficient as the hired worker; and that these results have useful

---

[9] Strictly speaking, the home produced activity of companionship can be provided by the market in the form of social escort services but to most people, this activity is considered non-substitutable for the simple reason that the utility of consumption of companionship provided by a non-household member would be substantially different from a household member (or marriage partner).

implications for households' decisions on whether to hire domestic help or use other market substitutes.

In valuing NMHP, the present study identifies two of its major components as home education, and family organisation and supervision in the home. A certain part of household production involves the teaching and helping of children with their school work, acquiring skills and giving parental advice on problems other than school work. The result of this activity is a form of human capital investment in the home. The amount and economic value of the time spent by parents with their children in human capital investment in the home may be a significant factor in a child's learning process.[10] While recognising the role performed by household members in providing home education to their children as part of household production, no attempt is made here to measure the extent or magnitude of the children's benefits from such human capital investment in the home for two reasons. First, the measurement of human capital investment is often tricky as the output or gains to the recipient is derived only in the long run and as such, difficult to apportion and ascertain. Second, to leave out human capital investment in the home is also consistent with the social accounting framework where the aspect of human capital investment in the market is ignored and only consumption expenditures on market education and training (expenditure approach) or factor payments to teachers and instructors (income approach) are included. Instead of measuring human capital investment in the home, in the present study, it is postulated that home education is positively related to the quantity of time spent by parents with their children and that a measure of the economic value of this time spent can be derived by using the average general market wage rate of kindergarten and primary school teachers. The work of these teachers come closest to being identified as a substitute for this household function, involving the home education of young children.

For the other major component of NMHP -- the time spent by household members on supervision and organisation of household activities (including supervision of domestic help, if any) and matters relating to paper work (settling bills, budgeting, etc) -- the study takes the average general wage rate of managers of very small firms in the market as an estimate of the economic value of family supervision and

---

[10]   Leibowitz, 1974 reported that one of the important findings of the Coleman Report was that by the time children enter their first grade in school, significant differences in verbal and mathematical competence exist among them.

6

organisation in the home. Although admittedly crude, the work of these small-firm managers come closest to being identified as part of the job requirements of a homemaker.

The study contends that the correct value of family household production for social accounting purposes requires the valuation at the margin of both MHP and NMHP. Traditionally, only the MHP is valued and the set of activities defined as NMHP are virtually left out. These NMHP activities constitute an important and observable feature of household formation and in principle, not substantially different from market provided services of education and tuition, and as part of the job requirements of managers and therefore capable of appropriate imputation.[11]

Using the model and method of valuation of MHP and NMHP suggested here, empirical estimates on the quantity and economic value of household production are derived for Singapore. The application of the model and method of valuation is particularly apt in the case of Singapore since domestic help services are easily available and fairly widespread. Further, these empirical results would be of interest for a variety of reasons. Apart from the relative scarcity of empirical estimates on home production in general, there has been to date, no work done on the non-market household economy of Singapore.

Like so many other countries, the national accounts of Singapore as well as other government publications such as the household expenditure survey, the manpower and labour force survey do not attempt to measure household production and continue to classify those engaged in unpaid household production as part of the economically 'inactive' population. Similarly, although there are several academic studies on the role and contribution of women to the Singapore economy, only minor and fairly general references are made in these studies to the role of women in the non-market household sector (Wong 1975, 1980; Lim 1982).

This study thus includes a first attempt to gather some information on the non-market household economy of Singapore. Apart from the attempt to measure and value household production in Singapore, the empirical estimates and analyses would also be useful for the purpose of answering the following questions:

---

[11] Note that market available domestic help services often do not include as part of their services the activities defined as NMHP and therefore to impute a value of NMHP, it would be inappropriate to use the remuneration of domestic help services.

1   How large is the value of home production in Singapore in relation to its current GNP as compared to some other industrialised countries?

2   What patterns of time use exist in Singapore and who are involved in household production?

3   What is the nature of the goods and services produced in the home?

4   To what extent has capital been substituted for labour in the production of goods and services at home?

5   How does the value of household production compare to the family's monetary income?

6   How does the value of home production differ among families with different geographical (locational) and socio-economic backgrounds?

7   Is there a relationship between the market labour supply of women and the value of household production and if so, what?

8   How does the value of household production using the method of valuation suggested here compares with the simple replacement cost and opportunity cost approaches using the empirical estimates derived for Singapore?

9   What public policies may be generated from having a knowledge of the value of household production in Singapore?

These are only some of the questions to which this study addresses itself. The study's objectives would have been fulfilled if it can be shown here that the methodological problems inherent in household production measurement and valuation are solvable and that it calls for a serious and concerted effort to institutionalize household production measurement on a regular basis. To the extent that the omission of the measurement and valuation of household production generates distorted information about the economy, the study would be of much value in correcting this imbalance.

8

The purpose of this work is to investigate the meaning and viability of measuring and valuing household production for use in social accounting. In addressing these objectives, it should also be made clear as to what issues the study is not principally concerned about nor able to address.

First, the study is necessarily economic in nature and approach. This involves postulates of maximising behaviour of households in their utility functions and the analysis use both economic monetary (e.g. income, wages) and non-monetary (e.g. family size, type of dwelling) variables. Sociological or psychological perspectives, though important in any study of family households are left out in much of the discussion that follows. This is primarily due to the author's lack of training and knowledge in these two fields. Thus, considerations such as stress factors in household production, intra-family interaction involving emotions and conjugal relations, and the such are deliberately but regrettably left out.

Second, no attempt is made to discuss household behaviour and the household economy using Marxist analyses or ideologies for the same reason as above. But more importantly, Marxist perspectives are rather different in emphasis, the focus being primarily on the macro production and reproduction of labour power and how society as a whole organises the production of goods and services. In the Marxist literature, the micro aspects of household behaviour and household production which this study emphasizes are virtually ignored.

Third, this study is concerned with the value of household work as done by the entire family and not just those done by wives or for that matter, by husbands. It is also important to recognise the distinction between valuing a wife's services and that of valuing a wife per se. The latter would involve more complex calculations -- if indeed they can be done -- and serious non-economic considerations are bound to arise. Studies purporting to value a wife as a person are thus highly suspect and suggest confusion in the objective of intention of measurement. The more meaningful and relatively easier estimate for the value of a wife or husband is derived from the services she or he performs and not of they as a person. Studies attempting to derive the former so as to assert the latter would remain meaningless efforts.

Fourth, while it is true that household production can be considered as part of a larger set of domestic or subsistence activities (say, in the economic development literature), no attempt is made to discuss the latter activities for the reason that they involve activities which are found in mostly less developed countries having a wide rural based economy, and thus non-relevant to the pursue of this study. Activities such as collecting firewood, helping out in the farm, cottage industries and handicrafts, among other activities are hence excluded.

Fifth, no allusions are made with regard to feminist literature. Indeed, feminist ideologies tend to shun women involved in household production calling the work demeaning and debasing.

Finally, the study is concerned with imputations for the value of unpaid household production within a social accounting framework. Here, the term 'social accounting' is used in a broad sense to denote first, a system of organised accounts showing economic transactions whether actual or imputed in an economy -- synonymous to national income accounting -- and second, the application of the information collected for analyses -- in this case, a study of the household economy involving for example, intra and inter household interacting variables affecting the determination of the amount and value of household production. What is not discussed however, is the intricacies behind the setting up of actual social accounts and the various sector accounts within that system of accounts. For they involve more complex elements and constitute by themselves an interesting but different study.

## 1.2    Organisation of the study

The study is organised as follows: Chapter 2 explains the need for the measurement of household production, specifically for social accounting purposes. It discusses the theoretical justification for household production measurement and the increasing trend of empirical investigations on the magnitudes of household production and its relationship with the market sector of the economy. The Chapter also discusses some of the likely obstacles and limitations of incorporating household production estimates in national accounts. The Chapter concludes that while household production measurement is necessary to understand structural changes in the economy and other intra-economic changes, it is likely to be less useful and may in fact be erroneous to compare household production estimates across national boundaries.

Chapter 3 deals with some of the methodological problems inherent in any research on household production; in particular, the quantification and the definition given to household production, the measurement of joint-production activities and the valuation of household production.

10

Chapter 4 surveys some of the major works and empirical estimates on household production valuation and quantification. Chapter 5 first presents a model of market replaceable household production (MHP) which focuses on a household's organisation of its consumption and production activities in the home and in the market in such a way that maximises the household's total utility. This theoretical model is then applied to national income accounting and the method of valuation of market replaceable household production derived from the model is shown to be consistent in the way market goods and services are valued within the national income accounting framework. Second, a discussion of the two major components of near-market replaceable household production (NMHP), namely home education and family supervision and organisation in the home is made. Here, the Chapter shows how these two home activities can be measured and valued.

In Chapter 6, the model and valuation of household production is applied to a household survey in Singapore to generate empirical estimates. Here the research design and survey questionnaire and procedures are described. The results of this exploratory survey on the demand for household production in Singapore and the factors which affect this demand are presented. With some simplifying assumptions and extrapolation, an attempt is made in the Chapter to calculate the value of household production in Singapore. Also, some comparisons are made between the findings -- the values or magnitudes -- of the present study and those of others using the replacement cost and opportunity cost methods.

Finally, in Chapter 7 some of the implications of household production estimates for such aspects as labour force participation rates of women, the ratio of household production value estimates to GNP and its possible trend over time, real economic growth with household production estimates included and excluded from current GNP, the distribution of real income and some policy issues on homemakers are discussed.

Chapter 7 will also provide some comments on possible implementation of the method of valuation and research design suggested here on a much wider scale. To inject some caution in interpreting these results, the concluding chapter also outlines some of the limitations of the present study including problems still unresolved and discussion on whether household production has been accurately imputed. Finally, and briefly discussed are possible future research priorities on household production.

# 2 The need for measurement of household production

## 2.1 The resistance to measure

As early as 1898, Marshall writing in his *Principles of Economics* correctly observed that "... a woman who makes her own clothes, or a man who digs his own garden or repairs his own house, is earning income just as would the dressmaker, gardener, or carpenter who might be hired to do the work."[1] But in his later 1910 edition on the discussion of social income, Marshall put forward the main reason as to the reluctance of national income accountants to include the contribution of housewives in the social accounts in that it was so difficult to estimate the value of that contribution.

Following Marshall, Pigou, writing in his book, *The Economics of Welfare* (1946) raised what has become a well-known paradox in national income accounting that:

> ... the services rendered by women enter into the dividend when they are rendered in exchange for wages, whether in factory or in the home, but do not enter into it when they are rendered by mothers and wives gratuitously to their own families. Thus, if a man marries his housekeeper or his cook, the national dividend is diminished.[2]

Although recognising the asymmetry of household production measurement in national income accounting -- the work done by domestic help included whilst the work done by housewives excluded -- Pigou finally

---

[1]    Marshall, 1898 pp. 149.

[2]    Pigou, 1946 pp. 33-34.

decided in favour of excluding the latter from the national accounts citing a number of intractable conceptual problems and inadequate statistical coverage and empirical work. This is largely because virtually no records are kept on how much is produced nor the time devoted to providing such services, and that household services are generally not purchased in market transaction with prices indicating values at the margin.

Similarly, Kuznets in his definitive *National Income and Its Composition* volume (1941) remarked that:

> Exclusion of the products of the family economy characteristic of all national income estimates, seriously limits their validity as measures of all scarce and disposable goods produced by the nation ... Over long periods distinct secular shifts occur in the relative contributions of the business and the family economy to the total of economic goods, most broadly defined. One must, therefore, guard against the common tendency to consider national income totals as all conclusive summaries of the scarce and disposable sources of satisfaction produced by the nation. Such summaries would become practicable only if the data improved substantially or if the family disappeared entirely as a producer of goods.[3]

However, Kuznets final decision was also to exclude household production from the social accounts. His decision was based partly on the lack of available data to implement the desired changes and partly on his intention to define a limit that only goods and services having market-based valuations be included in the GNP.

In principle, most economists recognise that household production has value in the same way as market production but point out the extreme difficulties in estimating such a value and with the lack of information and data on intra-household activity, they are prepared to exclude it from the national accounts altogether. But there are other more serious problems to household production measurement. These are the conceptual and methodological problems. Mentioned somewhat briefly in the previous chapter, these problems include definitional and quantification problems; method of accounting for joint household production, and the appropriate valuation method consistent with the social accounting framework.

Unlike market transacted goods and services where an observable exchange takes place, and prices generated, household production to a

---

3    Kuznets, 1941 pp. 10.

large extent remains a non-market activity with no observable prices and hence do not get counted in the country's market-based GNP. And, even if one should decide to include household production in the GNP, the problem of what exactly constitutes household production to enable data gathering purposes remains a problem in much need of a solution. To give an example, should an activity like taking a child for a walk in the park be considered as child care and hence part of household production or as leisure since the parent also enjoys the walk. Thus problems on definition of household production necessarily invoke the long standing debate among economists as to what activities are to be considered economic and thus capable of pecuniary imputation, and those that are non-economic. In other words where should one draw the line between economic and non-economic activities?

Given that the definitional issue is solvable, the next immediate problem concerns the quantification of housework. In other words, having identified what activities are to be included as household production, the problem of aggregation of the activities or the volume of output produced sets in. The nature of housework is complex and variable. Not only are there many activities performed in the home but also some of these activities can be performed simultaneously and/or by more than one household member raising such questions as how such work is to be reported and added.

The third methodological problem essentially asks what method of valuation should be used to value household production? There are currently two competing methods: output-related evaluation approach versus input-related evaluation approach. Each of these approaches has its own merit but the question before a national income accountant is which of these methods of valuation will produce the most meaningful and consistent set of estimates of the economic value of household work that is also consistent with the way income accountants measure market goods and services?

The solution to the preceding conceptual and methodological problems is crucial since they determine an accurate accounting of the amount of household production and its economic value. For unless there is general consensus among national income accountants, there would be no uniformity in data gathering and the estimates generated for household production would remain exploratory and highly unreliable, if not erroneous. Each of these problems is discussed in detail in Chapter 3 and some possible solutions offered.

Unless such methodological problems can be resolved and to the agreement of income accountants, the resistance to measure household production would remain strong and unfavourable.[4] In some of the attempts to impute values for household production, these methodological problems were either ignored or casually dealt with, resulting in questionable estimates. To include such a doubtful calculation and estimates in the GNP would greatly reduce the reliability of the total estimate of production and the reluctance to include this item of activities by income accountants is, thus, understandable. In 1930, King writing on the series of income studies, published by the Bureau of Economic Research says:

> ... the Bureau has naturally been compelled to confine its investigations to types of income translatable into terms of money units. Items upon which no money value can be placed have necessarily been omitted ... Among such omissions are the value of the services of persons to themselves and to their families ... The value of such home services is tremendous ... Yet because the difficulty of correctly evaluating such services is so very great, they have been excluded.[5]

Aside from these methodological problems and the acute scarcity of reliable data, opponents to measuring household production for social accounting purposes have also argued that because of the large dissimilarity between non-market household produced goods and services and market purchased goods and services, any effort or tendency to impute values on home production from market analogues would render such attempts meaningless and incorrect. Unlike household production, the existing imputations for a few non-market activities -- rental income on owner-occupied housing, farmer's consumption of their own food crops, etc -- are acceptable since these are cases for which very clear market equivalents are present.

---

4     Perhaps the strongest resistance to measuring household production lies in the anticipated high costs of changing the whole structure of the national income accounting framework to incorporate such a measure. Further, it has been argued that there would be no stopping here, since other non-market economic activities will inevitably demand representations in the GNP. The national income accounts, as presently constituted, are mainly used by businesses, and any suggestion for changes might be upsetting.

5     King, 1930 pp. 35.

Thus, for example, in the case of a farmer who produces food crops for sale in the market and also for own consumption, very clearly since in either case, the food crops are identical, then the imputation for that part of the food crops not sold on the market is a straight forward procedure; that being the market transacted price of the food crops. Such is not the case of household production; the arguments being (1) market purchased services, say, hired domestic help, do not carry an embodiment of love, care and a desire to promote the highest quality housework, and (2) that occupational choices and the allocation of time to housework and marketwork are dictated by the cultural norms of a particular society thereby making opportunity cost methods to valuing household production time and effort invalid. This also has the implication that a comparison of a modified GNP with household production included across different countries may be rendered meaningless when it comes to welfare comparisons.

In sum, the reluctance to measure and to include household production in the GNP rests on several valid grounds. That it is not desirable to include household production in the national accounts does not however preclude attempts to measure and value household production for other purposes nor even as a social welfare indicator synonymous to other social indicators (such as education and health). Since national income to a large extent represents a measure of income and consumption, it can be regarded as one of the social welfare indicators and not the only indicator, and that makes the valuation of household production all the more relevant. This issue is addressed in section 2.3.

Further, it appears that the resistance to measure household production for social accounting purposes is weakening since it can be observed that in recent years, there has been renewed interest in the subject in the form of published papers (see Chapters 3 and 4), seminars and conferences (NBER Conferences on Research in Income and Wealth, Household Studies Workshop in Asia, etc), more pages devoted to discussions on household production in both recent micro and macro textbooks, and policy discussions, among other recent developments. This rekindled interest is the result of many factors: increasing dissatisfactions with national income accounting rules; advances in economic theory; improvements in methodology of calculation; public interests in measures of quality of life; changing interface between the market and non-market sectors in both developed and developing economies involving increases in labour force participation of women, new technologies introduced in household production and the rising commercialisation of housework; and new discoveries for the use of empirical estimates on household production. This renewal of interest in household production research and the factors attributing to it is examined below.

## 2.2 Demands for measurement

The last two decades -- especially in the late sixties and early seventies -- saw increasing criticisms of the methods of national accounting and in the calculations of national income. These criticisms centre on three areas of contention: (1) inadequate measurement and erroneous designation of some national income components[6] (eg. consumer durable, education as investment and not consumption), (2) normatively incorrect objectives of the GNP[7] (eg. measures of economic welfare or quality of life arguments, social indicators movement), and (3) omission of non-market goods and services[8] (eg. volunteer work, household production).

The dissatisfaction with the GNP over the omission of goods and services not transacted in the market seems well-founded in the case of those goods and services having market equivalents and the characteristics of exchangeability. Thus, it is inconsistent of national income accounting rules to include say, domestic help services when they are rendered in exchange for money in the market and to exclude the same domestic help services when instead they are rendered without remuneration at home (the case of homemakers and their home production). Although a ready presumption is that services associated with the former would be inferior because of the lack of an ability to cater to the idiosyncrasies of members of a household and that of the affinity bonds of family members, the quality could also be greater. And, since present national income accounting rules do not include the value of such emotions or direct utility from the act of performing housework (and for that matter in market work as well) such considerations ought not to be of any concern.

The complaint of inconsistency in national income accounting rules also extends to present day imputations for some non-market goods while continuing to exclude household production. It is now widely accepted

---

6    See for example Hawrylyshyn, 1974; Juster, 1970.

7    Ibid.

8    Hawrylyshyn, 1974 and 1978; Carter 1979.

17

method of national accounting to include own-house rental imputations, and farmers' self-produced food consumption, both of which are non-marketed and have to be imputed. In this sense, it is considered inconsistent of national accounting rules to include some non-marketed goods and at the same time excluding others.

The objections raised that current imputed non-market goods have clear markets and identical market equivalents (in the form of goods or services transacted) unlike household production may not be totally correct say, in the case of imputing the rental income on owner occupied housing since we know that no property is completely alike. Further, it depends on the definition given to household production. Clearly a narrower definition such as market-replaceable household production would almost certainly ought to have very near if not, almost identical market equivalents. Such services as cleaning, laundry work and gardening, among others, are the same whether they are self-performed at home or purchased from the market. That is, it would be hard to put up a case for the existence of substantial differences between these market and non-market activities. For these reasons alone, such objections are not serious.

Thus, there is a continuing pressure for national income accountants not only to recognise household production as an economic activity but also to attempt some empirical measurements.[9] The idea is to have a set of imputations for non-market but economic activities supplementing but not adding to the market-based GNP for some of the varied uses for which figures of the former may be put.

Since Marshall, Pigou and Kuznets, the possibility of constructing a set of household accounts as part of an overall social accounting system has become more feasible given the improvements made in data availability and structural changes in most economies. Thus, for example, housework has become increasingly commercialised. The easy availability of domestic help services at affordable prices and the spread of commercial establishments such as launderettes, day-care centres and home cooked meal caterers have allowed more empirical data to be collected for analysis and imputation of the value of such services performed in the home but not exchanged for money.

The increasing availability of household data has also led to significant theoretical advances in microeconomics in the area of household behaviour and economics. Under the banner of the 'New Home

---

[9]  The subject of household production as an economic activity is discussed more fully in Chapter 5.

Economics', the household has been transformed from a sterile consumption unit to an active production unit engaged in the home production of utility yielding commodities, using market purchased goods and the time of family members as factor inputs (Becker 1965, 1976). Associated with names like Becker, Gronau, Schultz and Mincer, the 'New Home Economics' recognises that time is the more important input in household production as much of what is produced in households are services.[10] This, in turn, has led economists to focus largely on the question of the value of time spent in household production. For many reasons -- discussed in Chapter 3 -- the valuation of household output (in all its varieties) is very difficult and the valuation of household production time provides a more pragmatical solution.

Complementing the 'New Home Economics' are the increasing number of time-budget studies on households undertaken mainly by Cornell economists (Walker and Gauger, 1967 and 1973; Walker and Woods, 1976). These time-budget studies provide comprehensive accounts of the use of time by households on market activities, home production, and leisure. Commenting on the potential use of time-budget studies for valuing non-market household production, Nancy and Richard Ruggles (1970) wrote:

> The question of productive activity taking place within the household unquestionably needs further study. Time-budget studies on how people divide their total time among different activities (including eating, sleeping, and leisure) would be highly informative, and would provide a valuable set of data that could be directly related to the market transactions in the national economic accounts.[11]

Improvements in data availability and collection, and advances in microeconomic theory have thus provided the stimulus for further research on household production and a renewal of interest on studies on the household economy in general.

The need for the measurement and valuation of household production lies not only in a desire to provide a better estimate of the total economic production for an economy in a given year but also in providing a better

---

[10]   See Chapter 5 for an elaboration and formalisation of Becker's model.

[11]   Nancy and Richard Ruggles, 1970 pp. 40.

basis for inter-temporal growth calculations and for inter-spatial (international) comparisons of national income and product in both absolute and per capita terms. Thus, high growth rates as reflected in a country's GNP may be misleading as these may come from a diminishing household economy. It has been observed that over time, the labour force participation rate of women for most countries tends to rise[12] and there is, as mentioned earlier, an increasing commercialisation of housework activities so that, it is quite possible that total production of goods and services over time may not have changed that much.[13]

On the other hand, it might be argued that due to technological and social changes, demands for new goods and services have been created. One example is the increasing variety of market produced goods and services which in turn, makes household marketing and shopping -- a household production activity -- more time consuming than it once was. Another example, lies in the greater availability and use of the many types of mechanical household equipment (eg. blender, washing and dish machines, etc.) which allow more and in some cases better quality goods and services to be created (eg. more elaborate meals, more clothes washed). Thus, with increases in money income and changes in technology, society demands a higher standard of living and concomitantly,

---

[12]   In one 1984 report on women's progress, the labour force participation of women in the United States was shown as almost reaching 50 percent and rising and this may come from a diminishing household economy. It has been observed that over time, the labour force participation rate of women for most countries tends to rise and there is, as mentioned earlier, an increasing commercialisation of housework activities so that, it is reaching 50 percent and rising. The report concluded that "A major aspect of change has been the huge inflow of women into the U.S. labour force, a phenomenon which continues unabated" (Economics Road Maps, Nos 1976-1977, July 1984.)

[13]   As an illustration, take an economy at 2 points in time t and t+1. Suppose the total production (market and household) remains unchanged across the two periods at 100 units. Suppose in period t, the composition of production is 60:40 of market to household output and because of greater specialisation, the ratio becomes 70:30 in period t+1. This may be wrongly interpreted as an increase in GNP (60 to 70) if housework is excluded in GNP when in fact, production is unchanged for both periods. For some empirical estimates on the United States, see Maurice Weinrobe, 1974. By accounting for household production estimates over time, Weinrobe has shown that on an absolute level, the measured rate of growth of the GNP of the United States has been some 0.20 percent to 0.25 percent too large. And, on the basis of real output per potential worker, the growth rate has been overstated by some 10 percent.

more and better quality household production. This, implies that more rather than less household production has taken place over time.[14]

In sum, we do not know what the net result is given these changes on the rates of growth of both market and non-market produced goods and services over time but clearly it warrants investigation. Stressing the need for inter-temporal measurement of household production and on the implications of such a neglect, Weinrobe states:

> To the degree that production at home is not included in measured output, and to the degree that the omitted output changes over time in absolute amount and relative to actual output, both the quantity and trend of output are mismeasured.
> ... The problem which this fact entails is obvious. The rate of growth of measured output has been and continues to be a policy target. If the administrators of our economic society are going to guide their policy on a basis of such statistics then it is important that the statistics give good signals, and if the constituents of the society are going to judge the success or failure of economic policies then it is also vital that good information be available as the basis for such judgements. Our present information is less good than it might be.[15]

The value of household production derived for an economy and disaggregated by age of household members, size of household, income and other household characteristics would surely be useful guidelines for policy-makers in the event of a desire to influence the labour force participation rate of men and women in either direction -- that is, from non-market to market and vice-versa. In short, this leads to a better understanding of a country's labour supply. This, in turn, will allow policy-makers to come up with a better formulation of marginal income tax rates and observed changes in taxable revenue. For example, some empirical studies have shown that married women working outside the home react negatively to higher levels of taxation by reducing their paid market hours and increasing their time devoted to home production.[16]

---

14  An argument of this is found in Juster, 1973.

15  Weinrobe, 1974 pp. 90 and pp. 100.

16  See for example, Hunt et al. 1981; Rosen 1976; and Boskin 1974.

Indeed, one study by three University of Georgia economists concluded that:

> ... the hypothesis cannot be rejected that wives completely reallocate time lost from the labour market to non-market production in an attempt to restore household real income. Our results indicate that wives reduce labour supply but increase home production effort when faced with higher marginal tax rates.
>
> ... (But) the greatest loss to households from progressive taxation is an indirect one since households have the capacity to restore some lost income. The real loss may occur in the form of human capital depreciation of wives because of reduced labour-time attachment and lower hours of market work, an aspect of the tax structure that has been ignored.[17]

The information that married women's market labour supply reacts positively to net marginal earnings gives rise to several pertinent questions: what is the magnitude of the loss in welfare associated with increases in income taxation of married women to the household? How large is the withdrawal or injection of married women from or to the market following changes in marginal tax rates? And, is it likely that the tax system influences the economic role -- in market output production (in GNP and growth) and home production (provision of home goods and services, marriage patterns, and fertility behaviour) -- of all women in society? These are relevant questions in any study on labour supply of women and when properly answered would aid in decision making by policy-makers not only in economic issues but also of social ones.

By ignoring the non-market household economy, a measure of income distribution based only on market income is obviously deficient. In setting the poverty line level and a measure of welfare for example, a rather unacceptable proposition may arise that two households with similar characteristics -- size of household, location of house, number of children, type of house, etc -- and earning the same money income are equally poor or rich when one household has a full-time homemaker and the other does not. What is needed as such, to correct this misleading proposition on welfare is a more comprehensive measure of income distribution which will include the household's sources of non-monetary income as well. Thus, if a measure of the income distribution could be improved by

---

17    Hunt, J; DelLorme, C and Hill, R, 1981 pp. 446 and 452.

22

knowledge of an estimate of the value of household produced goods and services not included in household money income, then a better measure of the distribution of economic welfare could be devised for policy purposes.

Perhaps, one of the more important application of the measurement and valuation of household production lies in the measurement of economic welfare. It was the work of Nordhaus and Tobin (1972) in constructing a 'measure of economic welfare' (MEW) differing from current national income accounting rules that provided the impetus for further research on household production and welfare measurement. The issue here is whether the exclusion of household production in GNP makes the GNP statistic a biased indicator of changes in welfare.

It has been argued that because national income accounts include both income and consumption which are, in effect, welfare estimates, the GNP statistic can be taken to be an indicator of welfare.[18] But because, the GNP includes among other things, market goods and services that make no obvious contribution to individual economic well-being (eg. police firearms, anti-pollution equipment) and excludes items which contribute to economic welfare (eg. leisure, household services), the GNP statistic as it is argued, is a flawed index of a country's economic welfare.

Reconstructing the social accounts to generate a measure of economic welfare, Nordhaus and Tobin estimated the value of non-market household production in the United States to be about one-half to one-quarter of their MEW measurement in 1965. Further, Nordhaus and Tobin have shown that when such things as the value of leisure and household production are added to GNP and costs of pollution and other disamenities of modern urbanisation are taken out plus other adjustments, the result shows a positive growth in economic welfare (the adjusted GNP) but a slower rate of growth in the unadjusted GNP.[19]

On the same argument that there is a need for the measurement of household production, in that, GNP does not take into account of changes in economic welfare and indeed could lead to serious misinterpretation on the welfare of households, Bailey (1962) commented that:

---

[18]  See for example Studenski, 1958 pp. 3-5. Also see a discussion on GNP and welfare by Beckerman, 1976 pp. 38-61.

[19]  Later, the authors called this measure of economic welfare, 'New Economic Welfare' (NEW).

23

... if national income estimators use cash income as the sole indicator of the welfare of the household, serious and misleading interpretations of the nation's economic life result. The reason is that a rise in the amount of time devoted to leisure and unpaid economic work will reduce the time and income of paid employment, thus reflecting in the national income a decline in the welfare of the population. To correct this situation, a value should be imputed for leisure and for the unpaid employment of persons and resources ...[20]

Similarly Juster (1973) added:

In examining the relation between a system of social accounts and the measurement of economic and social performance, the question of what constitute feasible objectives of an accounting system must be kept in mind. We want the accounts to record changes in the material well-being of the community. That evidently means that the accounts must register changes in the flow of goods and services going through the market mechanism where the bulk of economic activity takes place, but it does not preclude the accounts from registering non-market activity to the extent that it bears directly and measurably on material well-being.[21]

Measuring household production also explains why some countries (especially less developed ones) with very low market GNP per capita can survive. In less developed countries, such omissions are substantial partly because their economies are less monetized and partly because there is inadequate data collection and thus information on such activities with the result that poverty tends to be exaggerated. Further, since the less developed countries are moving toward higher development stages, inevitably, there is a consequent rise in the commercialisation of housework and increased labour force participation of women, with the result that market-based GNP increases, although there might be no changes in the total physical quantity of goods and services available in the country. By not valuing household production, it then creates some distortion of a rapidly growing economy when it is not.

---

[20]   Bailey, 1962 pp. 282-294

[21]   Juster, 1973 pp. 26.

Of course this does not mean that 2 countries with the same GNP with household production included (GNP + HP) estimates enjoy the same level of welfare for two reasons: (1) the cultural norms with regard to market work and housework in the two countries may be substantially different. In the case of less developed countries, women are usually constrained by choices since in these societies, they are expected to be good homemakers. So that, given choices, women in less developed countries are more likely to choose less homework and more market work with the result that the value of their marginal product must be lower than say, women in advanced countries. Thus, we cannot say that although the two countries exhibit equal (GNP + HP) estimates, the welfare level is the same since one is constrained by choices and the other is not. And (2) GNP per se is generally not considered a true measure of society's welfare so that (GNP + HP) estimates are also not true measures of welfare.

Aside from the relevance of household production measurement for the question of GNP as a welfare and growth measure, labour supply analysis and the distribution of income, there are increasingly social pressures to recognise the economic contribution of women in households and in a society. These social pressures take the form of demands for a wider social security system encompassing housewives;[22] judicial recognition of women's contribution in say, the distribution of matrimonial property assets,[23] and in wrongful injury or death litigation.[24]

Given the various reasons and usefulness in studying household production and the household economy, it is time that due formal recognition and serious consideration be given to its measurement in social accounting. For if, macroeconomic textbooks continue to provide a footnote which admits household production has value but concludes that imputation is difficult and uncertain, and the fact that, national accounts and labour force surveys continue to classify those engaged in household production or homemakers[25] as 'economically inactive', then there can be no improvements or solutions forthcoming to these problems

---

22  On this aspect, see Proulx, 1978 Chapter. 5.

23  See Knetsch, 1984.

24  See Quah, 1987.

25  The term 'homemakers is usually taken to mean both housewives and house husbands ie. adult family members who contribute to household work in their own homes.

25

nor further contributions to knowledge in this area.

## 2.3 The desirability of a revision in the social accounting method involving household production

There are several aspects which call for readjustments in and extensions to the currently defined GNP statistic. Criticisms of the national income accounting rules on such things as failure to include or exclude some items from the income and product accounts, and the pursuit of the wrong objective of measurement, have always existed since the inception of the national or social accounts. The discussion here confines itself to some of the issues surrounding the proposal for including household production in the national accounts.

While proponents for the measurement of household production often point to the uses for which such measures can be put, opponents have argued that because of extreme difficulties in obtaining estimates and the high degree of unreliability often associated with past attempts, it would be preferable to exclude household production from the social accounts altogether. Over time however, the measurement of household production becomes more relevant due to changes in the social and economic environment, and this had led to increasing demands for income accountants to include such measures. This issue was discussed in the preceding section. In what follows below is a discussion of an issue basic to national income accounting: the fundamental notion of what constitutes an economic activity in social accounting and the question, on whether imputed household production estimates should be added to the conventional GNP.

*The meaning of economic activity in social accounting*

The need for social accounting scarcely requires justification. A nation or a society needs to know its economic performance and how it has utilised its scarce resources so that appropriate corrective policy measures, if required, can be undertaken. Thus, the first problem facing national income accountants is to define the boundary of economic activities. Specifically, the old and complex question of what is an economic activity?

Most broadly defined, any activity which uses up scarce resources in return for monetary and/or non-monetary satisfaction can be considered economic in nature. Clearly, however, using such a definition of economic activity would result in an inclusion of all human activities -- the universe of human actions -- in the social accounts. For this reason, and in order to make some meaningful sense, national income accountants have delimited all transactions that take place in the market as economic activities and all transactions outside the market as non-economic activities (with few exceptions).

The few non-market activities considered as economic activities and included in the national accounts are explicitly mentioned in the United Nations, 'A System of National Accounts' (SNA). Following the SNA, it appears that virtually all activities which result in the production of goods can be included in the GNP.[26]

Thus, activities like production of agricultural products for own consumption, self-made clothing, and the construction of own homes, all share the common characteristic that they do not enter into the market but produced for own (or family) consumption.[27] The values of these non-market goods are imputed from the prices of identical or similar goods sold in the market, and accordingly get counted in the GNP.

Guidelines for the treatment of production of services for own (or family) consumption are however conspicuously absent, perhaps the implicit instruction to ignore them. The only exceptions here and mentioned in the SNA are accountings for the rental income on owner-occupied housing and banking services. As noted earlier in this chapter, the treatment of the services provided by homemakers in the national accounts is asymmetrical. The convention is that if home services were purchased from the market, say, by the hiring of domestic help, they would be included; if these services were performed for and by the family household without pay, then they would be excluded. Thus, wash your own clothes, and the service is ignored; pay someone to wash your clothes, and income is recorded.[28] Clearly, this national income accounting rule on home services is not only arbitrary but also inconsistent. The usual justification for such an asymmetry in the treatment of housework is one of expediency since the valuation of such services can be most difficult

---

26    United Nations, 'A System of National Accounts', Studies in Methods Series No. w Rev. 3, New York 1968.

27    Ibid. See paragraphs 5.13 and 6.19 - 6.24.

28    This has resulted in a popular joke on national income accounting that if there is a desire to increase national income, then simply have all the homemakers remunerated.

(see Chapter 3).

The major difficulty lies in trying to distinguish between work activities and leisure activities since for some home activities as to some people, they can be treated as work, while for others, as leisure. Thus, is active baby-sitting to be considered as 'work' or 'leisure'? Too many activities in the home present such a problem and arguably, it is much easier to exclude them even if it means distortions in national income estimates, restrictions, and other often-heard criticisms.

It is perhaps obvious that what national income accountants had in mind in the design and purpose of the national accounts was a highly developed market-oriented monetary economy where in such an economy, the amount of activities performed in the home is negligible and rapidly declining. Thus, it is argued that no significant attention need be paid to the production in the household sector. This assumption is incorrect since studies on household production have generally indicated that an approximate magnitude of household production appears to be 30 percent of the GNP of such advanced countries as the United States and Canada.[29] Clearly then, the amount and value of household production are by no means small and unless this is recognised by income accountants, serious misleading inferences from conventional GNP statistics used by policy-makers in particular, will arise (discussed in preceding section).

---

[29] See for example Hawrylyshyn, 1978 and Murphy, 1978 and 1982. Also see Chapter 4.

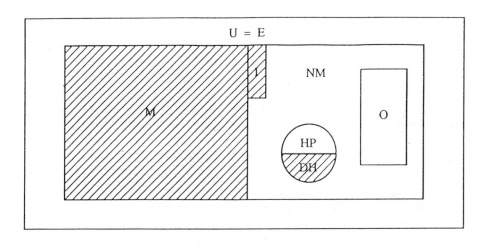

| U | = | Universe of Human Activities |
| E | = | Economic Activities |
| M | = | Market Activities |
| NM | = | Non-Market Activities |
| I | = | Imputed Non-Market Activities Included in GNP |
| HP | = | Unpaid Household Production |
| DH | = | Paid Household Production or Domestic Help Services |
| O | = | Other Non-Market Activities other than Household Production |

= Components of Measured GNP.

**Figure 2.1**  **Economic market activities, economic non-market activities and the GNP**

Figure 2.1 illustrates and summarizes the relationship among economic market activities, economic non-market activities and the GNP. In the above diagram, the Universe of human activities is taken to be equal to all economic activities most broadly defined (U=E). Within this realm of economic activities are those activities which take place in the market (M) and those outside the market (NM). In measuring the GNP, all market-based activities are included ie M⊂GNP. Then, there are those non-market but economic activities whose values are imputed and presently included in the GNP(I) ie I⊂GNP. Also, there are those household production activities which enter into the market in the form of paid domestic help services (DH) and as such are included in the GNP ie DH⊂GNP. Accordingly, the GNP as it is conventionally defined can be written as

$$GNP = M + I + DH$$

What remains within the realm of non-market activities and outside the GNP are activities like volunteer work, student work, etc. (classified as O in Figure 2.1), and unpaid household production (HP).

Given all the problems and objections mentioned earlier, should unpaid household production still be included in the GNP by way of imputations? There is no clear-cut answer to this question. But a reasonable answer would be to say that it all depends on the purposes for which the national accounting data is put to. Thus, if one is only interested in a measure of market activity, then the inclusion of household production estimates into the GNP would be detrimental since the latter, being primarily non-market, would distort the meaning and interpretation of the national accounting data.

Also, if one's objective is for the determination of the necessary tax rate increases on changes in government revenue to effect policies, then the question of whether to include or exclude household production from the GNP does not really matter much since real income -- non-market produced home goods and services -- are currently non-taxable.

Per contra, if the objective is to use national income estimates solely as an indicator of changes in social welfare, then the inclusion of household production is clearly warranted, however with one qualification. As with any other GNP omitted variable, if it can be shown that the change in the relative share of household production to total market production is small and insignificant, then again it does not matter much whether one should take into account of household production. But in order to show this to

30

be true, one still needs to measure household production and monitor its size and value over time. Thus, probes such as these would be most useful before undertaking any revisions of the social accounts.

It must be noted that there are other benefits in measuring household production other than for national income accounting purposes (see section 2.2, this chapter). Without being repetitive, it suffices to say that the measurement and valuation of household production are and will be important given changes in the social and economic environment. The next chapter shows that perhaps some of the methodological problems usually associated with household production may not be that insurmountable after all.

# 3 Problems in household production research

The main obstacles to an accurate accounting of the amount and economic value of household production are, first, that virtually no records are kept on how much is being produced; and second, that the amounts demanded by households are generally not purchased in market transactions where prices would then indicate their value at the margin. This is largely because household production is normally provided by members of the same household that also consume the goods and services produced. While the recent trend of household production research seems to indicate some general consensus as to the solution of the first problem, namely the measurement of the quantity of household production using time surveys and time budget studies,[1] there is still no convergence to any generally accepted method of estimation of the economic value of household production.[2]

Of course, the lack of a market mechanism and market price in no way makes household production any less economically significant.

It only means that household production research necessitates its imputation through some indirect methods. However the success of these

---

[1]    Among the major works on time budget studies include those of Szalai et al., 1972; Walker and Woods, 1976; and Juster et al. 1978.

[2]    While there might be some who would argue that the application of the direct market replacement cost method of valuing household production has reached some general consensus among researchers, this author would argue otherwise. The reason is that, while a number of studies favoured the direct market replacement cost method, there are equally found a number of studies favouring the opportunity cost of time method so that the issue of finding a generally accepted method of valuing household production is far from being settled. See Section 3.4 on 'Valuation Problems and Consistency with Economic Theory' and Chapter 4 on 'Survey of Major Works and Empirical Estimates on Household Production'.

indirect methods in producing meaningful estimates will depend on, among other things, the quantification of household production, the method of dealing with joint production activities; and the method's consistency with economic theory. But first, consider the immediate problem of defining household production.

## 3.1 Definition of household production

Problems of definition add difficulties to obtaining accurate assessments. Even with careful precautions in survey design, as in time-budget studies, concerning what is and is not to be included, people's responses with respect to time spent and the economic value of household production can deviate widely depending on how they perceive, the different household activities. For some, cooking, general household repairs and gardening may be viewed as pleasures or hobbies, while others look on them as differing little from cleaning, ironing and after-meal clean up. In other words, how do we distinguish between household work and leisure?

A number of economists associated with the New Home Economics, including Becker, are prepared to define household production in terms of the household commodities produced.[3] This, however, does not solve the problem since the notion of household commodities within the context of the New Home Economics is too all-encompassing. The definition of a household commodity in the New Home Economics is one which is produced by using time of family members and market goods; the consumption of which yields utility to the household. No attempt is made to distinguish between work and leisure. As these household commodities can be anything ranging from cooked meals to religious accomplishments and even sleeping (to use some of Becker's examples), it implies that virtually nothing produced by a household's non-wage labour that yields utility to the household can be excluded, clearly an unsatisfactory definition. But, of course, Becker and the New Home economists were more interested in the theoretical modelling of the allocation of time in home production rather than on the aspect of actual valuation of the household commodity produced. A more practical definition must be sought.

One involves the setting up of time budget studies which list the various categories of household activities that are considered by the household

---

[3]    Becker, 1965.

researchers as having satisfied their definition of housework.[4] These time budget studies involve the use of a modified diary which details account of all the time devoted to each household production activity over a 24 hour time-period and usually the recording is kept over several days. Although, all of these time budget studies try to be as comprehensive as possible in their coverage of the variety of household tasks performed, it is quite possible that some activities may be left out for those studies without a proper definition of household work. The lack or absence of a clear definition of household production leaves the household members unable to distinguish appropriately household work from leisure for those activities which were not on the given list of household tasks. Worse, the inclusion of a 'miscellaneous' category often results in the large number of hours reported (Hall and Schroeder, 1970).

In some studies, the definition given to household production may be too vague to be of any meaningful use and in some cases, again very large amounts of time devoted to household production were reported by households. The famous Walker and Woods time budget studies on housework for example, defined household production as those 'purposeful activities performed in individual households to create the goods and services that make it possible for a family to function as a family'[5] and a study involving nine international researchers defined household production as those 'activities performed for and by household member(s) that result in household production, the goods and services that enable a household to function.'[6]

Attempts by other authors to define household production also run into problems, Gronau, for example, defines it as similar to market production, in that work at home carries the same disutility (utility) as that of work in the market. His argument runs as follows:

> An intuitive distinction between work at home (i.e., home production time) and leisure (i.e., home consumption time) is that work at home (like work in the market) is something one would rather have somebody else do for one (if the cost were low enough), while it would be impossible to enjoy leisure through a

---

4    Supra note 1.

5    Walker and Woods, 1976.

6    Cited from Walker and Woods, 1976 Chapter 9, pp. 261.

surrogate.[7]

The problem with Gronau's definition is that it implicitly assumes that work, whether performed in the market or in the home, is unpleasant, whereas in reality, people do enjoy at least some aspects of their jobs and experience the relative differences in utilities and disutilities between jobs.[8] Despite the shortcoming, Gronau's definition is certainly an improvement over Becker's as it makes possible distinguishing production from consumption (or leisure) activities in the home.

Based on the same idea is Hawrylyshyn's Third Person Criterion of household production. According to this criterion, household work is defined as

> Those economic services produced in the household and outside the market, but which could be produced by a third-person hired on the market without changing their utility to the members of the household.[9]

To use Hawrylyshyn's example, the satisfaction of utility that a household gets from having a clean floor is not much reduced by the fact that work was done by another person outside of the household. The same, however, cannot be said of attending a symphony concert as the household cannot truly benefit (in terms of utility gain) from not attending the concert but paying someone outside of the household to attend on its behalf. Hawrylyshyn's definition therefore makes it possible to account for all production activities in the home which have market substitutes. It is, however, customary to exclude conjugal relations except in litigation over the value of a spouse.

Both Hawrylyshyn's and Gronau's definitions of household production place emphasis on whether activities, normally performed by a household member can be performed by someone outside of the household. While recognised as imperfect, as the definition may not cover all the household tasks performed by households, it is certainly an improvement over those definitions that are ambiguous.

---

[7]  Gronau, 1977, pp. 1104.

[8]  Admitting this, Gronau was quick to point out that 'even casual observations would indicate that this assumption is wrong.' See Gronau, 1974, pp. 635.

[9]  Hawrylyshyn, 1978, pp. 19.

The idea of using the criterion of whether a household task can be performed by someone outside of the household for it to be included as household production is not new and goes back more than fifty years ago when Reid writing in her book, *Economics of Household Production* (1934) stated that

> Household production consists of those unpaid activities which are carried on, by and for the members, which activities might be replaced by market goods, or paid services if circumstances such as income, market conditions, and personal inclinations permit the service being delegated to someone outside the household group.[10]

Thus, Reid's definition limits household production to those household activities capable of being performed by hired-help or at least carry a market equivalent.

Reid's definition of household production has however been criticised for providing a relatively limited job description for the homemaker (Hefferan 1982; Proulx 1978). For there are some household produced goods and services which are not market replaceable; such activities as the love and care shown to family members are not easily found in market equivalents. While this remains a problem, for most purposes requiring a measure of the economic value of household production, a more useful and functional definition of household production is the more crucial. The problem is that, if the definition given to household production is one which is broad and which tries to encompass every household activity, then the meaning and measuring of household production according to some criterion and deterministic mechanism of valuation is lost. The point is that, we need to define household production in such a way as to make it specific enough so as to enable data collection which is meaningful and for which the task of valuing household production for a specific objective can be undertaken.

Along the same lines as Reid, Gronau and Hawrylyshyn is the definition of household production offered by Beutler and Owen (1980):

> Household production is by and for household members and is market replaceable in the sense that it can conceivably be

---

[10]    Reid, 1934, pp. 11.

delegated to a paid worker.[11]

This does not suggest that household members do not engage in other activities, but definitions based on the notion that household work can be done by someone else does offer a means for clear focus on the production that is to be valued rather than on consumption that is not.

So far, we have argued that household production must be clearly defined in order to produce meaningful estimates. The task of finding a suitable definition of household production would, in turn, greatly depend on the specific objective of measurement and valuation of household production. There is no one nor should there be one definition of household production for all uses.

Thus, for example, if the objective of valuation of household production is for national income accounting purposes, then to be consistent with the valuation of market-based activities, the definition of household production should be one that is capable of market valuation. On the other hand, if the valuation of household production is for own-family welfare considerations, then the definition of household production would have to be more inclusive (that is, broad) other than merely being capable of market valuation. Such things as conjugal relations, love and care for family members then become important and should form part of the definition of household production.

In Chapter 1, household production has been defined as consisting of three types namely, market replaceable household production (MHP), near-market replaceable household production (NMHP), and non-replaceable household production. But only two types of household activities -- MHP and NMHP -- should arguably be used as the definition of household production if the objective of measurement and valuation is for social accounting purposes.

Recall that market replaceable household production are those unpaid home activities which could be done to prescribed specifications and to the benefit of the household by someone outside of the household, usually through purchased domestic help services. Examples of such activities are cleaning, washing laundry, etc. Near-market replaceable household production are those unpaid home activities which do not find easy and ready replacements in the market but yet could conceivably be done by appropriate specialised hired-help. Examples of such activities are tutoring a child, some managerial and supervisory roles. These activities

---

11   Beutler and Owen, 1980, pp. 18.

are not usually associated with domestic help services. Finally, non-replaceable household production comprise those home activities which are not replaceable by any hired-help or use of market substitutes. Examples of NHP are the love, care and companionship among household members.

In imputing an economic value for household production, consistent with the way marketed goods and services are valued i.e. social accounting, clearly only those home activities which have market or near-market equivalents - MHP and NMHP - are included. The value of household production can be seen as being derived from having it done and not by having it done by any particular individual or member of the family. Except for possible quality and efficiency differences and, of course abstracting from any consumption benefits in the act of production, the value of the output, and therefore of the benefits is not attributable to the input of any particular individual or individuals.

It must be emphasized that in social accounting, the valuation of household production should be based on the viewpoint of society or what outside impersonal forces look at. In this case, it is the work done or service provided and not the embodiment of love that should matter.

Thus, the definitions of household production proposed and used by Gronau, Reid, Hawrylyshyn, Beutler and Owen can be seen not only as attempts to facilitate measurement and valuation but also are the theoretically correct definitions of household production if the purpose is for social accounting. However, their definitions are still too narrow for they exclude to a large part the NMHP activities. On the other hand, the definitions of household production utilised by Becker and supported by Proulx and Hefferan are much too broad for they include MHP, NMHP and NHP -- all in one definition. A clear definition of what constitutes household production is required to avoid any ambiguity, concerning what is meant by the term 'household work' and if any meaningful set of estimates of the quantity and economic value of household production are to be derived.

Having defined household production, the next problem is that of quantifying those home activities which fall within this definition, this being the subject of the next section.

38

## 3.2    Quantification of household production

The non-market nature and variety of household production make its measurement difficult but challenging. For a start, although individual home produced goods and services can be identified and in some cases quantified, such quantification however can only be in terms of specific goods and services as the number of clothing items washed, the number of meals served, the number of made beds or the number of times a household went grocery shopping per time period. Although strictly speaking, it is possible to conduct a comprehensive survey itemising all the goods and services produced by households, such surveys would be extremely expensive to operate and would be immensely cumbersome in view of the detailed information required.[12]    Thus for all practical purposes, it is not possible to obtain an aggregate of the physical goods and services produced nor is it possible to quantify one item in terms of another.

The time spent on household production by household members on the other hand, presents a reasonably good alternative to the measurement of household production. The method of using time spent in performing the varied household chores as a measure of household production has roots in traditional home economics but recently given more attention in the United States by the numerous time use surveys and time budget studies undertaken by Cornell University, New York State of Human Ecology and the Institute for Social Research, University of Michigan. The most extensive research ever done on time use is the 1964 UNESCO sponsored multinational study involving twelve countries. This multinational study provided much useful information on time use in the various countries on

---

[12]    One very exploratory study involved an estimate of the value of food production in households based on a survey by comparable food served in the market (restaurants, etc.). While one can imagine the extreme difficulty in getting information on the types of food served by and for household members due to the wide variety possible, even more difficulty would be encountered when other household outputs are included. Such things as clean floor, disciplined child and after meal clean up are not quantifiable and are not directly purchased from the market making valuation impossible. See the study on food valuation by Stafford and Sanik, 1981 for a taste of the difficulty involved by using this method of household output valuation.

leisure, paid employment and unpaid household work.[13]

The immediate advantage of using time as a unit of measurement of household production is that since all household production activities require time, a unit of account in the form of total time used can be established. For example, time spent in cooking, cleaning, grocery shopping and in other household chores can be added up to yield the total time spent in household production. Moreover, not only is time additive but also it is expressible in different units such as an hour, a day or a week so that data collection in whatever form is made easier. Using time as a measure of household production also allows us to determine the average cost of production for which wage rates can be abscribed to in units of time (say $X per hour); thus commensurating with wages earned in other kinds of employment.

Using time to measure the work load in household production has however two major problems. First, the range over which time spent in household production will be an accurate index of the quantity of household output produced will depend on diminishing returns. If diminishing returns to labour does occur and is very high at low levels of time spent in household production, then the range over which the index is accurate will be narrow. So that, beyond some initial hours spent in household production where there exists constant productivity, an additional unit of time spent may not yield the same output produced as the initial hours. This problem becomes more severe if diminishing marginal productivity occurs early.[14]

Second, household production requires not only labour but also non-labour resources and given household technology in production. Market goods are bought and combined with the time of household members to produce household goods and services. For example, a clean floor is produced by soap-detergents bought from the market and the expenditure of time of household members in cleaning the floor. Similarly, a home-made cake is produced by the combination of cake making utensils (mixer, trays, etc), cake materials (baking powder, sugar, etc) and the time of the person making the cake. The point is that, should there be wide variations in the use of non-labour resources and changes in household technology, then using time as an index of measurement of

---

[13] For a report on women's time and expenditure in this study, see Szalai, 1975 in Futures pp. 385-399.

[14] On the same note, see Bryant, 1982, pp. 2.

40

household production may be inaccurate since with better household technology and/or change in the labour-non-labour combination of inputs in household production may require less time to produce the same output as before. This problem becomes more severe over long periods of time. Here a good example is that of the time spent in washing clothes. With the advent of washing machines, the time spent in washing clothes today is much lesser than years before for the same load of clothes that needs washing. But this does not mean that less clean clothes are being produced by households today. Note however, that if the purpose of measurement and valuation is for national income accounting, then this problem becomes less significant as then time is used as a measurement of value-added of the household sector. Thus, time used in household production is only an approximate index of household production with diminishing accuracy in the long run.

Perhaps, one ought to regard household production not in terms of physical outputs but rather as an activity or service provided by household members for which the final output of the activity takes the form of physical units of a home produced good. Thus, the time spent in cleaning for example, gives us an indication of the amount of cleaning activities performed in the home. Although, still subject to the same problems when using time as a unit of measurement of the physical quantity of household production, at least it now focuses on the amount of household production services provided within households which in turn are clearly capable of being valued. Thus, while it is difficult to place values on clean floors and made beds, it is a much easier task to value the services of cleaning and making beds which are services commonly found in the market and therefore marketable (more on this aspect of valuation later).

Further, since efficiency differences in production occur across different households -- the same amount of household output produced with some households using less time -- there is no reason to suspect that an accounting of the average time spent in providing household production services for an economy will be biased in one direction. Also, one can compute a fairly accurate average measure of the time spent in household production activities for households having similar household characteristics or family composition (income, size of family, age of family members, etc). In fact, one major finding of the well-known time budget study of Walker and Woods (1976) revealed that:

The most important result of the study has been the confirmation of a direct relationship between family composition and time spent on household work, thus allowing the use of time spent on the work to become a measure of household production.[15]

Similarly, the study by Hawrylyshyn (1978) for Statistics Canada concluded that the number of hours spent on household work appears to be a fundamental indicator of household work.[16]

In sum, although theoretically, either amount of the household work accomplished in terms of physical units of a particular home output or the time expended to perform household work (given some assumptions) can be used to quantify household production, it is in measuring the time input in relation to the physical output that allows a more practical or useable measure. Thus, despite some of the problems mentioned earlier, household production can be seen in terms of household production activities or services requiring time inputs for which the output -- household produced good -- can also be measured in terms of time, albeit imperfectly.

But there is another problem in the use of time as a measure of the quantity of household production. This is the problem of joint production.

**3.3      Measurement of joint production activities**

A third major problem in household production research concerns the measurement of multiple activities which occur simultaneously within the household. Some tasks, such as laundry and caring for children can often be done concurrently with other activities, raising ambiguities on how such time is to be reported.[17]

Even when household members are able to give an accurate description

---

[15]   Walker and Woods, 1976, pp. 262.

[16]   Hawrylyshyn, 1978, pp. 6.

[17]   A study of simultaneous activities by Steeves et al. (1981) for example showed that of all household production activities, shopping for daily necessities and groceries was most likely to be done as a separate activity whilst food preparation, house cleaning, family care and travel were likely to be done with another activity. However, the study also reported that when the activities were done simultaneously, they were more likely to be accompanied by interactive family care and other personal interactions. This seems to imply that where simultaneous activities do occur, a great deal involves non-replaceable household production which is not easily measured or valued.

42

of the nature of the household tasks involved, how do researchers weigh the relative significance of each of the tasks? This question is exigent when it comes to the valuation aspect of household production. Thus, if it is the case that both dish-washing and the care of children occur simultaneously, then should the time spent and its economic value be attributed to dish-washing or to child care or both?

One method commonly used is simply to allow for these simultaneous activities, in that, time is counted twice for any two activities performed simultaneously. Similarly, if a household member performs three activities together, then the time common to all three activities is multiplied by three.

A major problem with such a method is that it can lead to a gross exaggeration of the total amount of time spent performing household work. It is thus quite possible for households to report that they had spent more than 168 hours a week performing household chores. Such overstatements of time devoted to household work usually arise when it is the case that while performing one activity, the individual gives only intermittent attention to the other. Thus, for example, a household member may be doing some cooking, while occasionally glancing over or tending a child. The latter activity could hardly be given the same amount of time as the former.

A more accurate method of accounting for time use in households where joint production activities are concerned is to ascribe the time entirely to the major task. In other words, time is recorded only for the major household task performed and the other activity is ignored. Thus whether the time use involves two or three simultaneous activities, it is left to the household reporting the hours spent on household production to decide which is the major activity. This eliminates the need for researchers to weigh the relative importance of each of the tasks. Further, a special commissioned study on simultaneous activities in households by Steeves et al. (1981) for the Family Economics Research Group of the U.S. Department of Agriculture reported in its findings that where simultaneous activities are involved, a large part concerns non-replaceable household production such as interactive family care and other personal interactions.[18] Recording for time use only for the major task involved appears to be a better solution. At least, the method works within the framework of a 168-hour week constraint, so that a micro-allocation of time for market work, home work and leisure, can be derived and the

---

[18]    Ibid.

implications of such an allocation of time for the different households easily drawn.

## 3.4 Valuation problems and consistency with economic theory

Valuation methods that will produce meaningful estimates (i.e., consistent with economic theory) comprise the fourth major problem in household production research. The problem arises because household production activities occur outside the market with the result that there are no prices to indicate the value of such outputs. Economic valuations, of necessity, are dependent on indirect means of imputing value measures. Just because this production is an extra-market activity does not imply that the goods and services produced have no value. Households clearly value them and demonstrate this by willingly giving up other goods and services in order to enjoy the benefits of their provision.

The problem, however, is not whether household production has value or no value, but rather is one of the appropriate method of valuation. The appropriate method will, in turn, depend critically on the purpose of the valuation: whether it is for national income accounting, matrimonial property settlements or the valuation for compensation questions. Furthermore, the most common methods of valuation -- replacement cost and foregone wages -- are not always appropriate as will be shown below.

### 3.4.1 Existing methods of valuation

There are essentially two general approaches to valuing household production. One, the expenditure approach values the total output produced by the household in terms of what the output could be sold for in the market. The second approach, called the income approach values household production in terms of its resource inputs. While the expenditure approach is still very exploratory and suffers from serious operational difficulties, the income approach appears to be more favourable as it is empirically applicable. There are still a few variations of these two general approaches. Here, we identify four of them, namely, the reservation wage, Gronau's marginal productivity method, Pyun's utility income method and Colin Clark's institutional replacement method. Each valuation approach and its variants are discussed below.

*The expenditure approach : Output-related evaluations*

Modern economic theory views the family-household as a firm engaged in the production of utility-yielding home goods and services using market purchased goods and the time of family members as factor inputs.[19] In searching for the similarities between production in firms and in households, some researchers have argued that the valuation of household output can be derived from the market valuation of the firms' output. Called the expenditure or product-accounting approach, the method asserts that the value of household production can be estimated by summing up the prices of market goods and services comparable to those produced at home.[20] The main advantage of such a method is that the expenditure approach measures quantity of household outputs directly.

As an illustration, the expenditure approach was recently revived by a French economist by the name of Goldschmidt-Clermont (1983) who had published two papers on the conceptual advantages of using the approach and proposed several steps for which the value of household production could be derived.[21] Following are her evaluation procedures:

(1)     Select a household commodity that is market replaceable and determine its resource costs in terms of inputs of time and money in producing that household commodity.

(2)     Find a market price and the amount of time the household needs to acquire it.

(3)     Compute the difference in money costs between producing the commodity at home and consuming from the market.[22] The net monetary savings (or expenditure) yields the value to be assigned

---

[19]    Supra note 3.

[20]    Goldschmidt-Clermont, 1983; Stafford and Sanik, 1981, 1983; Chaput-Auquier, 1959 among others.

[21]    Goldschmidt-Clermont, Dec. 1983, Sept. 1983.

[22]    These money costs, say, for home laundering, include such things as cost of utilities (water, electricity, gas, etc) and amortization and repair of washing machine and the costs incurred in sending one's clothes to commercial laundering.

to household production. This is also the value added resulting from household production.

(4)     Compute the net time cost between household production of the commodity and household consumption of the time taken to acquire the market equivalent.

(5)     Divide (3) by (4) to arrive at the household wage per hour (or per unit time).

(6)     Compare the quality of the household produced commodity and the market equivalent. Should there be quality differences that are not expressible in money terms (i.e. economic terms), these household products are to be acknowledged explicitly.

While Goldsmith-Clermont offered no empirical work on her evaluation procedure, two Ohio State University researchers, Stafford and Sanik (1983) applied the expenditure approach to household food production to derive the value of home production. Similarly, other empirical studies adopting the expenditure approach are found in Suviranta and Mynttinen (1980: unpaid laundry work and cooking), and, Suviranta and Heinonen (1979: home care of children). These studies are very exploratory and restrict themselves to a selected sample of households. Overall, studies using the expenditure approach are very few and for these few studies, only easily identifiable household outputs, say, food and clean laundry, which have close market equivalents are valued. Implicitly, the expenditure approach entails major and seemingly insurmountable problems in application.

The output produced by a firm is valued by its quantity times the price at which it is sold. The price of each unit output of a firm is market determined in that it is established through the interaction of the forces of demand and supply. These prices and the units of output produced by a firm have been agreed upon by buyers and sellers, and are observable.

Obviously, the expenditure approach to valuing household production will be very difficult to operationalise in the case of the family-household. First, there are no common units of measurement of the total quantity of household outputs and no market price to reflect the economic value of the amount of household output produced. Second, even if we can quantify the amount of household production, say, in terms of units of output for a specific home good such as the number of meals served, clothing items made, and value them at market prices of similar meals and clothing products sold in the market, there are still many products produced at home which are not sold on the market and therefore no market prices to indicate values. Even such common household produced goods as clean floors, made beds, disciplined child and many other tangible and intangible products have no market equivalents to establish values.

Some defenders of this approach have suggested that in the case of home goods which have no market equivalent in the sense of tangible outputs, say clean floors and made beds, a measure of the value for these goods can be derived from the wages paid to the services of hired domestic help.[23] This is theoretically incorrect. First, many of the other home goods having market equivalents, say, cooked meals can also be produced by the hired domestic help, thereby, making it inconsistent to value some of the home goods in terms of output-related evaluations and others with input related-evaluations. Second, the suggestion is also inconsistent with national income accounting in that one do not mix an income approach with an expenditure approach in deriving the GNP. Clearly the wages of domestic help represent a payment to labour and should not be taken or added to the market equivalent value of another household output, the latter which is an expenditure item.

The third problem of the expenditure approach is that, for the expenditure approach to be fully operational for those household produced goods which are quantifiable in units of a specific output, say, for the number of meals served, then to arrive at a fairly accurate imputation of the value of home cooked food would necessitate very detailed and fairly accurate knowledge of what kinds of meals were served, the ingredients which go into making the meal, the cost of the utilities incurred in cooking such as gas or electricity charges and the number of times this particular meal has been served. Also, information is required of similar meals being sold in the market. Obviously, the method is very cumbersome and moreover, this information is required for all or at least an appropriate size of households, making data collection an immense task. For all these reasons, the expenditure approach is an unsatisfactory method by which to value household production.

The second approach to valuing household production, namely, the income approach appears to be more applicable to the household since many of the resource inputs are purchased directly from the market or at

---

[23] This suggestion is made by Goldschmidt-Clermont, 1983. Supra note 21.

least capable of market price imputations.

*The income approach: Input-related evaluations*

The difficulty and complexity of operationalising the expenditure approach to valuing household production have led researchers to seek alternative methods. By far, the most commonly used method is called the income or factor payments approach. Essentially, the method sums up the values of each of the resource inputs used in household production just as one aggregates the payments to factors of production to obtain the national income. In the case of market production by firms, the income approach would involve summing such things as the wage paid to labour, rental payment on land, input cost of raw materials, interest on capital borrowed and corporate profits before taxes to yield the value of the firm's production. In the case of the household, this would mean summing the values of the goods and services sold in the market and purchased by households (e.g. raw materials like meat and vegetable; services from consumer durable like refrigerator and blenders; and other goods like cleaning detergents), and the time spent by household members in production of household commodities (e.g. cooking and meal preparation) as well as time expended in supervising, organising and budgeting of household work. Some of the household commodities produced are the washed clothes, clean floors, cooked meals, and well disciplined and tutored children.

However, just as there are problems in valuing market production using an output-related evaluation approach, the same problems also arise in valuing household production using this approach. For example, in valuing consumer durable such as household appliances used in household production which are several years old, is the value to be assigned be based on its original purchase price of its value today? Similarly, the house in which home production takes place may have been purchased for $100,000 ten years ago, what is the value of the house today and how much of this was used in household production? Further, how should one value the time spent by family members in household production?

These problems are indeed complex ones but fortunately they can be answered in relation to the purpose of the valuation. In national income accounting for example, consumer durable like cars and refrigerators are treated as nondurables such that consumer expenditures on them are counted only once and that is, in the year they are made. In the case of housing services, the national income accounts treat the owner-occupied home as a business, with the housing service as the product which the owner buys from himself. The value-added by residential housing is derived by the existing rental market of similar type of housing. Thus, with regard to owner-occupied homes yielding a flow of housing services, a measure of rental income is already included in the GNP.

By far, the most conspicuous omission of household returns to factor inputs is the value added of time spent by household members in household production. Unlike market work where time spent is remunerated and included in the calculation of national income, there are no wages for home work and consequently no values assigned to the time spent by household members engaged in household production.

Thus, many researchers who have used the input-related evaluation approach to valuing household production had quite correctly focused solely on the question of valuing household production time. While market purchased goods are important inputs in household production just as time is, they are already captured in the GNP the moment they leave the factory. Adding them to household production valuation for a second time would result in double counting.

With regard to capital inputs such as cars and household appliances, there are good reasons for ignoring the value of their contribution to household production. Apart from the fact that they are already counted in the GNP via market transactions, there are extreme difficulties in establishing an appropriate economic return to capital. The family car, for example, could be used to transport groceries and bringing children to school which are part of household production but the same car could be used to provide a leisurely drive around the countryside or seeing a movie, thus making it very difficult to determine that part of capital that is used solely for household production and consequently the value added. Further, there is the difficulty in establishing depreciation allowances for capital goods used in household production. How does one, for example, obtain an estimate of depreciation of a several-year old oven?

Thus, given these complex problems and the fact that most non-labour inputs of household production are already counted once in the GNP, a measure of the value added of household production should be in terms of the valuation of time. Also, since most measures of homemaking are based on time spent in household production, valuation usually involves the application of a wage rate to the time devoted to producing household services. There are essentially two general approaches used to determine this wage rate. Time may be valued in terms of the wage rate paid to market substitute labour or by an estimate of the income or wage rate foregone by not devoting the time spent in household production to paid market labour. Each of these two approaches to the valuation of time spent in household production is discussed in turn.

*Replacement value estimates*

The most popular method used to value household production performed by members of a household is to equate this value to the cost that would be necessary to hire another person or persons to do the same work.[24] This method, termed the replacement cost or value approach has two variations.

The first applies an appropriate replacement wage rate to the average amount of time that is devoted to each of the different types of activity at home. The total time taken up in household production can be broken down into separate time use for such activities as meal preparation, laundry, child care, cleaning and so on. The total value is then taken to be the sum of the hours devoted to meal preparation multiplied by the market wage rate of a cook, plus the hours devoted to cleaning multiplied by the market wage rate of a cleaning lady, plus the respective hours taken up by other household tasks, each multiplied by the wages commanded by people who hire themselves out to do those sorts of jobs. Proponents of the method argued that, while some household tasks do not lend themselves to easy classification in terms of employment or job categories, most do and approximations are usually made for the rest.[25]

The second variant of the replacement value method evaluates what it would cost to replace all household services by hiring a single maid or housekeeper.[26] Rather than assuming that different tasks would be done by cooks, babysitters, chauffeurs, cleaning maids, and the like, who, it is sometimes argued, would demand premium wages to work the intermittent hours that most households usually give to such activities, this approach makes the more realistic presumption that most of the household work would be done by a single individual characteristic of households and who quite often performs several households activities simultaneously. The replacement cost by housekeeper approach is thus more appropriate than

---

[24] Hawrylyshyn 1978, Walker and Gauger 1980, Peskins 1982 among others.

[25] In other words, the basic assumption here is that household work has close counterparts in the market.

[26] This method is suggested by among others, Yale 1982, and Rosen 1974.

its alternative variant, the replacement cost by specialised function, in that, it takes the wages of the generalist domestic help rather than those of the specialists to valuing household production, thereby recognising the similarity to the generalist homemaker.

There are however, a number of criticisms levied against both approaches. Either versions of the replacement value method has the inherent weakness that both quality and efficiency between home producers and market workers are assumed constant. Quite apart from issues of having work done by a member or members of a household who, presumably, has/have affinity bonds with others in the household (which is quite another consideration to that of valuing household production) the quality of the work done by a substitute or a replacement person may be substantially different to that normally performed. Although a ready presumption is that it would be inferior because of the lack of an ability to cater to idiosyncrasies of members of a household, the quality could also be greater.

While there is not much that one can do to measure or adjust for quality differences between home produced and purchased services or for that matter even in market services per se, it is perhaps best to leave this emotionally charged issue alone as it is highly unlikely to be resolved satisfactorily.[27] The suggestion by some authors to add or reduce by some X percent of the replacement value estimates is untenable as it is not only arbitrary but also very subjective to be of any meaningful use.

But note however that if the purpose of the valuation of household production is for national income accounting, then this aspect of quality differentials need not be of much concern since the valuation of household production is from the viewpoint of society. The relevant questions are whether such services are available in the market and whether they could

---

[27]  It is often argued by some writers that the homemaker will produce greater quality work than hired-help simply because of greater love and care among family members. Clearly there is some truth in this for some but not all cases. Some studies have shown that households with women employed in the labour market nonetheless spend many hours in housework and that for those households which hire domestic help, they do so for those household chores for which they believe they have the least relative advantage. Further, it might be argued that for those households which chose to enter the labour market rather than working full time at home did so because they considered that to be their best option, while for those people who chose to be domestic helpers did so because they believed that was where their relative advantage lay. Thus, on average there is no a priori reason to expect an upward or downward bias in the quality of hired workers.

be done to prescribed specifications by the employer.

With regard to efficiency differences between homemakers and hired-help, the argument is that if household members are more efficient in performing housework than the market substitute labour, then, the replacement cost method will underestimate the value of household production. Conversely, if household members are less efficient in performing housework than the substitute labour, then, the replacement cost method will over estimate the value of household production. Most, if not all, studies using the replacement cost method do not take this efficiency differences into account in that they do not attempt to adjust their resulting replacement value estimates. The present study departs from these previous attempts by imputing an average relative efficiency coefficient between home produced and market purchased services to adjust the replacement value estimates derived (see Chapter 5).

Another commonly heard criticism of this popularly prescribed replacement cost method to valuing household production, is that, the method essentially excludes household management and organisation, which are, important elements of household production. Thus, even if the household makes use of hired help, a household member must manage or supervise the hired labour in organising household work, budgeting and other aspects of planning. Consequently, it has been argued that the replacement cost estimates undervalue household production by excluding this important function of the household. The present study however attempts a measurement and a valuation of these household management activities (see Chapter 5).

Still another criticism of the replacement cost method is that, if the household values the replacement services at market cost rates, then it is argued that the household would have purchased them rather than producing them at home.[28] That for the most part, household services are not purchased implies that households value these services less than the market value or that the act of production -- performing the household service -- yields utility to the household member. Consequently, it is argued that the replacement cost method overestimates the value of household production in the case of the former and underestimates in the case of the latter.

There is some validity to this criticism of the replacement cost method but again it must be noted that the criticism only holds when one is valuing household production strictly for household welfare purposes; that

---

[28] See for example, Gronau 1980, pp. 413 and, Zick and Bryant 1983 pp. 135.

is, on the implicit value of household production from the perspective of the household. If the valuation of household production is however for social accounting purposes, then it is through the perspective of the market for these kinds of services or that of society's valuation that becomes relevant. Thus, for example, a person may value his teaching profession more highly than is currently reflected in his wage rates but if society values it differently, then, it is his wage rates and not his own valuation that indicates the market valuation of his service. Further, it might be noted that in social income accounting, no attention is paid as to whether the producer of a service derives pleasure in the act of production. Only his or her market wages count.

Finally, there are some other minor problems with the replacement cost method to valuing household production, but these are particular to the replacement cost by specialized function method. One involves the argument that when the specialized function approach is used, full-time wage rates of specialists are applied instead of part-time ones. Since almost all household tasks are performed only part-time, and that part-time wages are generally lower than full-time wage rates, there is a tendency for the value of household production to be overestimated using such a method. Also, it has been argued that the services performed in the market are in the presence of very different capital goods than those present when households tasks are performed at home raising problems of incompatibility between market work and homework.[29]

In sum, despite some of the problems mentioned above, the replacement cost method to valuing household production does have conceptual and pragmatic advantages. This is particularly so in the case of the replacement cost by housekeeper approach. To recapitulate two of its major advantages:

(i)    A large part of household production consists of daily household chores which are market replaceable. Clearly, home services like cooking, cleaning, laundry and after-meal clean up, among other services, can easily be performed by hired-help. There are little or no special skills involved such as the requirement of certificates or other professional and tertiary education. And since, hired-help are themselves currently engaged or have been engaged in household production for their own family, the paid household

---

[29]    Note that these problems do not arise in the case of the replacement cost by housekeeper approach.

work that they perform for other households should not be substantially different from theirs. The point is that, the nature of most household tasks performed at home should not be so different from those performed by hired-help as to render the market services of hired-help incomparable to those performed by household members at home.

However, the argument that not all home services are market replaceable is correct. This brings to mind of activities which comprise of non-replaceable household production such as the love and care among household members, conjugal relations and other activities requiring more personal attention. But clearly, these activities are also excluded from possible imputations for social accounting as they involve to a large extent much broader and possibly, non-economic issues. Again, unless one is interested in measuring the economic value of the household's own evaluation of household production taking into account of the love and care (the so-called 'personal touch' or 'embodiment') of household functions, then the inclusion of non-replaceable household production in valuation violates the social accounting framework.

There are, however, a group of activities produced in the home by and for household members which are not normally associated with paid domestic help. In the present study, these activities are labelled as 'near-market replaceable household production' and comprise of two major activities: the home education of children and household management (see Chapter 1, Section 1.1.1). Thus, using the market wage rates of domestic help to account for these activities would be incorrect since they do not normally form part of the services of domestic help. An alternative method of valuing these activities is required. This alternative is shown in Chapter 5 of this study.

(ii) Unlike the replacement cost by specialised function where the method suffers from the difficulty of selecting the appropriate market equivalent occupations for each individual household function, the replacement cost by housekeeper approach makes the more realistic case that normally most of the household chores are done by a single generalist household member. The serious problem of accounting for simultaneous activities also disappears since the hired housekeeper can also perform several household tasks within the same hour.

A frequently proposed alternative to the replacement cost method is to value the home production time in terms of foregone money income. This is called the opportunity cost method.

*Opportunity cost estimates*

An alternative suggested means of valuing the time devoted to household production is to abscribe to this time the opportunity cost of the income that is given up when time is devoted to household services rather than earning money income from paid employment outside the home.[30] The method assumes that the rational individual will have allocated time to household work so that at the margin the value equals the forgone market wage rate. The choice of allocating time to household production rather than to wage employment is taken to imply that the household values the household work more than the money given up, and therefore, the opportunity wage income forgone is then a measure of the value of the household services. Further, because the housework is chosen in preference to the money, it is arguably a lower bound measure.[31] The estimate of the value of the time people spend performing housework is calculated by using the market wage rates for those who are currently employed while for those who are currently unemployed, their potential market wages are utilized.[32]

---

[30] Among the proponents of the opportunity cost method are: Nordhaus and Tobin 1972, Komesar 1974, Posner 1977 and Pottick 1978.

[31] Weinrobe for example argues that "the wife's decision to remain outside the market labour force reveals that she values her time at home at least as equal to what she could earn in the market place." See Weinrobe 1974, pp. 89. The same argument is made by Bryant when he wrote that 'since the full time housewife is not employed, then she and her family must implicitly value her time more highly than the families of otherwise identical married women who are employed, otherwise she too would be employed.' See Bryant 1982, pp. 7.

[32] Applications of the opportunity cost approach have invariably involved three variations, namely, (1) using average market wage rates, (2) after-tax market wage rates or take home pay and (3) net return to market work or net compensation. The first variant of the opportunity cost method makes use of an average of market wages plus any money supplements (see Nordhaus and Tobin 1972). The second variant takes the market wage rates and adjusts them for marginal income tax rates (see Sirageldin 1964). Finally, the third and more comprehensive variation of the opportunity cost approach requires the adjustment of individuals'
(continued...)

There are a number of problems that may arise in the application of such a procedure. One concerns the wage rates that should be assumed to be forgone. This may be a particular problem with individuals who have been out of the employment market for a long time and presently have little in the way of marketable skills. Another concerns the institutional constraints dictating the length of work weeks (such as 44 or 40 hours per week) thereby introducing a large element of 'lumpiness' to the options facing individuals. Because of this, it is not clear how much of the household production time is valued more than the outside wage and how much it is not -- even though the total wages are attributed to the sum of the work performed. In effect, each individual is faced with an all or nothing choice. The person must accept or reject a contract that specifies both the wage rate and required hours of work. Consequently, if individuals are working fewer hours than they wish to work and cannot find more market work, then the use of their current market wage rates would overestimate the value of their time spent in housework. On the other hand, for individuals who are working more hours than they wish to work, then their market wage rates underestimate the values of their time in household work. The problem becomes even more severe in the case of people who are involuntarily unemployed. In such cases involving people with no market opportunities, the use of the opportunity cost estimate would result in a zero estimate for the value of time spent in household production. Thus, given this 'lumpiness' problem, it may be inappropriate to value all the hours spent in household production at the market wage rate.

A more serious difficulty with the opportunity wage approach and in fact a source of much confusion is that the method implies that the value of household production increases the higher the education of household members performing the homework. An example would be a homemaker who is trained as a clerk earning $4 an hour vis-a-vis another homemaker who is trained as a medical doctor earning $12 an hour; the opportunity cost method would imply that the latter's household work is worth at least three times more than the former![33] The point is that the forgone

---

[32](...continued)
    market wage rates not only for marginal income tax rates but also for work-related costs such as commuting to work expenses and costs of eating lunch at work, among other things. See Murphy 1979.

[33]  Note that the method would also imply that Barbara Streisand's value of household production must be worth millions of dollars annually.

income might have no correlation to the skills required by a homemaker. To reconcile this problem, it has been suggested that it is not the value of the output produced as implied by the opportunity cost approach but rather the value of the homemaker's time or the value of simply 'being at home.'[34]

There are still many other problems with the opportunity cost method to valuing household production. But these problems need not be of any concern to us as our objective here is on the valuation of household production for social accounting purposes. Thus, problems such as the opportunity cost estimate, is, but, a marginal concept so that the infra-marginal units of time is worth more than the market wage rate and that there are disutility and utility in market work as well as in household work, are more relevant to studies involving the valuation of household production for individual's welfare purposes.[35]

We now turn our attention to other prescribed methods of valuing household production. Almost invariably, they involve a variation or combination of the output and input related methods of evaluation.

*Other methods of valuation*

*(1)    The reservation wage measure*

The reservation wage method was first proposed by Zick and Bryant (1983). Essentially, the method calls for the estimation of the minimum wage rate that would induce an individual such as a full-time homemaker to accept market employment. The idea is that, people choose to become full-time homemakers because they value their time spent at home to be greater than their potential market wage rates. As long as their potential market wage rates are lower than their own perceived values of home time, they will choose to remain outside paid employment. Raising their potential market wage rates to equal the values they place on their home time, will result in the homemakers being indifferent between accepting paid employment and remaining at home. When their potential market wage rates are above their perceived values of home time, they will be induced to accept paid employment. Thus, the reservation wage rate tells

---

[34]    See Firebaugh and Deacon 1980, pp. 60.

[35]    For a discussion on these problems, see Murphy 1978, 1982 and, Ferber and Birubaum 1980.

us the lowest potential market wage rate that will induce a full-time homemaker to switch from unpaid non-market work to paid market work. It is argued that the value of time spent at home equals this reservation wage rate.[36]

The reservation wage method suffers from three major problems. First, it is not a measure of the value of household production but rather a measure of the implicit value of home time placed by household members. Second, the values placed by homemakers on their time spent at home would necessarily involve consideration of leisure as well as the production of services and output. The reservation wage measure would thus incorporate an imputed value of home leisure as well as a value of time spent in performing housework. Third, even if the reservation wage method purports to measure the value of non-market time devoted to household work, the estimate derived from using such a method would have an upward bias. This is because the reservation wage would include not only fixed costs of employment but also work-related costs. In other words, the reservation wage includes such things as costs of commuting to work, payment of annual union dues, income taxes and other related costs of employment so that it overestimates the value of household work time. The magnitude of such an overestimation would be higher the higher are these fixed costs of employment and other work-related costs.

Although the reservation wage method is not appropriate to the valuation of household production for social accounting purposes, it has relevance for research on the allocation of time between paid and unpaid work, and like the opportunity cost approach, it provides a measure of

---

[36] As the reservation wages of individuals are not directly observable, recourse to a carefully designed survey questionnaire to elicit people's reservation wages must be taken. An alternative method to obtaining empirical estimates of the reservation wages has been used by Zick and Bryant (1983 pp. 137-138) involving what is now known as a Heckman econometric technique (Heckman 1976). Basically it involves the estimation of three equations:

$$W = X + e_1$$
$$W^* = Z + H + e_2$$
$$H = 1 (X - Z + e^1 - e^2)$$

where $W$ = the market wage rate, $X$ = a vector of market productivity determinants, $W^*$ = the reservation wage, $Z$ = a vector of home productivity determinants and $H$ = the hours spent in market work. For more details and empirical results using such a method, see Zick and Bryant 1973, pp. 138-143 and Heckman 1979, pp. 153-161.

what has been given up when time is devoted to non-market work and leisure. Also, the method may be useful for studying labour force participation of both women and men and their implications for labour and manpower policy considerations.

## (2)    *Gronau's marginal productivity function approach*

In his 1980 paper published in the *Review of Economics and Statistics*, Gronau attempted to measure productivity and the value of home production.[37] Using an innovative method to overcome many of the operational problems usually associated with the output-related evaluation approach, Gronau measured the value of home output by directly specifying the equation of the marginal home productivity function, $f_H$. His purpose is to derive the value of home production by the integration of $f_H$.

To estimate the marginal productivity function, $f_H$, Gronau suggests two alternative ways: one is to assume a certain functional form of the home production function, f (e.g. Cobb-Douglas or CES), derive $f_H$ and estimate it empirically or alternatively specify the functional form $f_H$ and integrate to derive f. In Gronau's paper, he chose the second approach by directly specifying the marginal home productivity function to be of the semi-log form, that is

$$\ln f_H = \alpha_0 - \alpha_1 H + \alpha_2 Y$$

where Y is a vector of socio-economic variables affecting the value of marginal productivity at home, and H denotes time inputs (also work at home). The set of explanatory socio-economic variables, Y includes such things as the wife's age and education, the husband's education and wage rate, the family's non-earned income, the number of children, the age of the youngest child and the number of rooms in the house. Using this marginal home productivity function, $f_H$ and noting the equilibrium conditions for utility maximization that is

$$f_H = S = W$$

where S = the marginal rate of substitution between leisure and goods

---

[37]    See Gronau 1980, pp. 408-416.  For an earlier paper of Gronau's more sophisticated modelling of the value of a housewife's time, see Gronau 1973, pp. 534-651.

$$MR_{LX} = (\partial U/\partial L)/(\partial U/\partial X) = S$$

where X = market goods and home goods; W = market wage rate; and L = Leisure,

Gronau derives the work at home function, H for labour force participants as

$$H = (\alpha_0 - \ln W + \alpha_2 Y)/\alpha_1$$

and this can be estimated as

$$H = a_0 - a_1 \ln W + a_2 Y$$

where $\alpha_0 = a_0/a_1$, $\alpha_1 = 1/a$ and $\alpha_2 = a_2/a_1$. With estimates of $\alpha_i$

where i = 1, 2, 3, the value of home produced goods (Z) can be obtained by the integration of $f_H$. That is,

$$Z = \int_0^H f_H (t)\ dt$$
$$= \int_0^H e^{(\alpha_o - \alpha_1 t + \alpha_2 Y)}\ dt$$
$$= e^{(\alpha_0 + \alpha_2 Y)1 - e^{(-\alpha. H)}} \frac{1}{\alpha}$$

The structural parameters of Z were then derived by using the restricted simultaneous equation method of estimation.

The marginal productivity function approach is still at its exploratory stage and apart from Gronau's, there has been no further work using such an approach.[38] However, one serious problem which may arise in the use of such a procedure is that if the objective of valuation of household production is for social accounting, then to estimate the value added of the labour inputs (or alternatively, the value of time spent in performing

---

[38] A similar household production function approach was proposed by Hawrylyshyn. In Hawrylyshyn's proposal, the value-added, V, is given by V = Vk + Vt and the gross value-added, Vg is Vg = V + Vi = Vt + Vk + Vi where Vk = value of household capital services, Vt = value of household time spent in housework, and Vi = value of intermediate purchases. Thus, only if data are available for each of these variables, the method is essentially impractical. See Hawrylyshyn 1977, pp. 79-96.

household services) contributed by household members would require a substraction of the value of capital inputs and other market goods from Gronau's gross value of home output. In other words, provision must be made for an estimate of some proportion of intermediate purchases, the magnitude of which to be deducted from the gross value. This estimation of the value of intermediate goods will be very difficult if not impossible for all practical reasons. As Gronau himself readily admits, 'to obtain estimates of the value added ... one has to subtract the value of market inputs. These inputs are, however, unknown.[39] Further, to account for these market inputs, Gronau proposes as part of his concluding comments that estimates of the value of labour inputs to home production can be obtained from the forgone wages of household members engaged in household production.[40] But this brings us back to the opportunity cost method of valuing the value-added of household production and renders the marginal productivity function approach as no different from the opportunity cost approach.

Leaving aside empirical problems, there seems to be no theoretical justification in explicitly selecting the semi-log functional form as representing the household's marginal productivity function, apart from saying that tests indicate that this semi-log variety produces better results than the linear and double log form.[41] Despite these problems, Gronau does succeed in providing an indirect approach to the valuation of home output and thereby overcoming a major problem usually associated with the output-related evaluation approach, that is, the undefinable home output.

*(3)    Pyun's utility income method*

Lamenting that most economic analysis supplied by economists on the problem of establishing the monetary value of a housewife have been 'flimsy', Pyun, a professor of economics at Mercer University, U.S., published a paper on the subject and offered his alternative based on a combination of opportunity cost-replacement value method. Although Pyun originally intended that his proposed method be used to determine

---

[39]    Gronau 1980, pp. 414.

[40]    Ibid., pp. 414.

[41]    Ibid., pp. 409, footnote 2.

the monetary value of household production as contributed by the housewife and therefore appropriate for courts dealing with wrongful death litigation, his model and method of valuation has since then been cited in a wide variety of works on valuing household production.[42] The method is discussed here only because serious flaws exist in both his exposition and methodology which have led to several incorrect assertions.

Figure 3.1 below reproduces the diagram Pyun uses in his article. Pyun

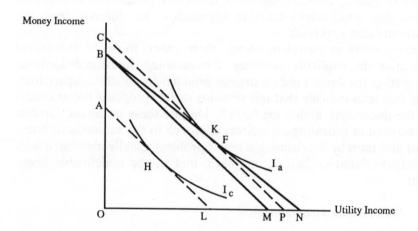

**Figure 3.1        Pyun's utility income approach**

42    Pyun 1969. Cited as one of the methods on valuing household services in Rosen 1974, Peterson 1978 and referred to in Murphy 1980, and Goldschmiodt-Clermont 1982.

assumes that a housewife can contribute money income (market production) and utility income (home production) to her household with the latter defined as "the housewife's utility creating capacity expressed in terms of the monetary value of her services computed on the basis of going wage rates for her various services in the household."[43] In Figure 3.1, this utility creating capacity of the housewife is measured along the horizontal axis (X) and is shown by the amount LN. Pyun takes the case of a housewife who does not work in the market and assumes that if she had worked, then the maximum amount of her monetary contribution is AB measured along the vertical axis (Y) and is equal to

$$X = \frac{1}{n} \sum_{i=1}^{n} X_i$$

where X is the prospective earnings of the housewife in the base year, $X_i$ is the average annual earnings of the ith occupation which she can take up and n is the number of all possible occupations which she can fill.

Pyun argues that the housewife's utility creating capacity to her household must be greater than the opportunity cost or wage income she could have earned in the market (her monetary contribution to her household) or otherwise she would have been employed. Thus LN > AB. He defines the indifference curves $I_a$ and $I_c$ as 'the behaviour line of the household as a whole toward the total utility derived from the combination of the X and Y income sources', and the household's income constraint line as BN (Figure 3.1).[44]

The household's initial equilibrium position is at point K, the point of tangency between $I_a$ and BN. If the housewife dies, the household will be on a lower in difference curve, $I_c$, indicating lower total utility with optimality is now at point H. Following Hicks, the substitution effect is derived by drawing a line parallel to AL and shifting that line until it is tangent to indifference curve $I_a$. He measures PN as the amount of the substitution effect and concludes that the loss to the household from household production is the amount LP.[45]

---

[43]   Pyun 1969, pp. 278.

[44]   Ibid, pp. 278

[45]   In measuring LP, Pyun provides a hypothetical case of a deceased housewife and uses a combination of geometry applied to his indifference curve diagram, and
(continued...)

A number of criticisms can be levelled against Pyun's utility income approach. Most of these criticisms arise from his incorrect and improper use of indifference curve analysis which has led to several incorrect conclusions. First, the household's element of choice between market work and home work or some combination of both, basic to indifference curve analysis, is absent as Pyun from the beginning considers the housewife as not working at all. Further this assumption is inconsistent with Pyun's own indifference curve diagram (Figure 3.1). In his diagram, the household's equilibrium position is shown to be at position F which implies a level of utility income that is slightly below OL, but yet we are informed that the housewife's utility income contribution is LN.

It is not clear from his analysis as to what Pyun means by 'utility income.' Pyun, by asserting that utility income provided by the husband, OL, is limited to the amount of his money income, OA, such that OL = OA would seem to imply a one to one transformation between money income and utility income, which then calls into question the validity of his income constraint lines and the shape of his indifference curves.[46] If OA money income provides OL utility income, then the line AL is not an income constraint line and the position H shown in his diagram becomes impossible to interpret. This assertion that utility income is perfectly transferable into money income, both in dollar terms, would necessarily imply that the two goods -- market and home produced -- are perfect substitutes and as such would require the indifference curves to be straight lines and not convex as he has drawn them. In essence, Pyun's income constraint lines are his indifference curves!

Since the objective of his paper is to determine the monetary value of household production as contributed by the housewife, it is ironic that Pyun starts off his analysis by assuming a replacement wage approach, as apparently evident by his so-called 'utility income creating capacity' measured on the horizontal axis, X, where LN represents 'the monetary value of the housewife's services computed on the basis of going wage

---

[45](...continued)
    statistical analysis to derive some estimates of the monetary value of a housewife's household production contribution to her family. See Pyun pp. 280-283, Statistical Appendix.

[46]    Pyun pp. 278. Also, see footnote 22 on the same page. The same criticism was brought up by Rosen 1974.

rates for her various services in the household.'[47]

Pyun's treatment of utility is also incorrect. Since the level of utility attained by an individual from the consumption of goods is reflected in positions of the indifference curves on an indifference map which is drawn in commodity space, Pyun, by measuring utility on the horizontal axis is destroying the foundation of the ordinal indifference curve analysis he seems to employ. He should have set up his analysis in such a way as to provide for a choice between money income and household production, or some other alternative on the horizontal axis. Then utility can be measured by the resulting indifference curve reached.

In sum, Pyun's utility income approach is seriously in error in its use of indifference curve analysis, although the same could not be said of his attempt and purpose. Given the serious flaws in Pyun's theoretical construct, what follows as Pyun's application of geometry to his indifference curve diagram, and his elaborate statistical analysis applied to a hypothetical example involving the loss of a housewife to a household, becomes irrelevant and as such will not be discussed here. While it was not Pyun's intention that the method he had proposed be used to value household production other than for compensation in wrongful death litigation, mention of his method (without much criticism) has been found in a number of studies involving the valuation of a housewife's contribution to household production.[48] It is introduced here not only because it represents one variation of methods used in valuing household production but also there is a need to expose some of the major errors associated with the method, which until now has been readily accepted by some works in the literature and cited as an alternative method of valuation. It might also be pointed out that despite the theoretical flaws, Pyun's utility income approach is totally inappropriate to value household production for social accounting purposes[49] for two main reasons: (1) it has, as its unit of analysis, only the housewife whereas the appropriate unit should be the household, and (2) it is concerned solely with deriving the compensating variation in income for loss of household services provided by the deceased housewife and hence more of an individual welfare valuation objective rather than as the perspective of a society's valuation

---

47  Ibid., pp. 278. Also, see footnote 21 on the same page.

48  Supra note 41.

49  To be fair, Pyun had not intended it earlier.

of household production.

## (4)    Colin Clark's institution replacement cost method

A variation of the replacement cost method is Colin Clark's proposal to value household production based on the cost of maintaining welfare institutions such as adults in old folks homes and children in institutional homes.[50] Statistics were obtained for state-run institutions and the total cost of upkeep for adults and children while deducting from this the total cost of purchased goods and services. The statistics were compiled in such a way that they exclude such items as the cost of buildings, rent and capital charges, and on expenditures on food, clothing, durable household goods (furniture), etc. The result is an estimated value of household production on a per capita basis.

While essentially a replacement cost approach, Colin Clark's institutional replacement cost method suffers from one major problem which may render the method inappropriate for valuing household production. The method makes the assumption that institutions and private households are the same in the nature of the provision of services and their costs of production. Clearly, this underlying assumption is incorrect. First, there might be significant economies of scale in large institutions which would then lower the cost of services provided to their residents.[51] Second, it is generally true that the quality and amount of services provided to residents in institutions will not be the same as those provided at home; it is more than likely the quality and amount of services provided in the former will be less than that at the latter because of the larger number of people being cared for in welfare homes and institutions.

What has often been overlooked or underemphasized in research on household production valuation is that the valuation may be based on marginal values, total values or net values depending on the purpose of the valuation. Empirical estimates on the value of household production have in themselves very little meaning if no particular attention is paid to the valuation question as to which is the theoretically appropriate method of valuation. The answer to this question lies in which type of economic

---

50    Clark 1958.

51    On the other hand, significant diseconomies of scale and other institutional inefficiencies may be present which then increase the cost of production of services per capital.

values is one looking for, and this, in turn, will depend critically on the purpose of valuation. This is the subject we examine next.

## 3.4.2 Distinguishing different economic values and their uses[52]

The peculiarity of the provision of household production is that to a large extent, households themselves supply them. The economic value of the time spent in household production by household members provides a reasonably good indication of the economic value of the benefits produced -- household commodities -- using that time. An indication of the expected gains and losses of varying levels of production effort to a household is illustrated in Figure 3.2.

Figure 3.2 shows the marginal benefit (MB) -- the incremental benefit from an additional hour devoted to home production -- and marginal cost (MC) -- the additional cost incurred from devoting an additional hour to home production -- curves (assumed here for convenience, linear and continuous) of a given household. The principle of declining MB indicates that for most households, at least some level of household services are important, but successive quantities are likely to be less important. At point N, nothing could be gained by devoting an additional unit of time to home production. The MC curve is shown to have a positive slope indicating higher additional costs as more time is employed in household production. For most households, the MCs are likely to be relatively small for some limited allocations of time to home production, and may even be zero or negative for some very small amount of time (OJ) but would probably increase for larger commitments. This rise in MC corresponding to increased household time spent in home production is the result of putting the more market productive household members with respect to the use of their time to household production.[53] A household facing MC and MB as indicated in Figure 3.2, would devote OR hours to household production. For by devoting OR hours to home production, the welfare of the household is maximised.

We can identify and measure the different economic values of time

---

52  This section draws mainly from my published papers in the *American Journal of Economics and Sociology* (Quah 1986a) and *Applied Economics* (Quah 1987). Also, see Murphy (1982).

53  It should be pointed out that this reason cited for diminishing returns to household production is only one (and not the only) way of rationalizing this phenomenon.

devoted to home production depending on the purpose of measurement. First, there is the notion of marginal value which is the value which households placed on the last unit of time (say hour) to home production. In the case of Figure 3.2, this marginal value is represented by OW for the ORth hour of home production. Second, there is the notion of total value which is the sum of all the hours performed in home production. In terms of Figure 3.2, this is shown as the area OVPR for OR hours of home production. Finally, there is a residual value, more commonly called the net value from home production. By net value is meant the residual benefit from the total value over total costs of home production. Since the total cost of home production is given by the area under the MC curve i.e. JPR for OR hours and the total value is OVPR, the net value measurement is then OVPJ(OVPR - JPR).

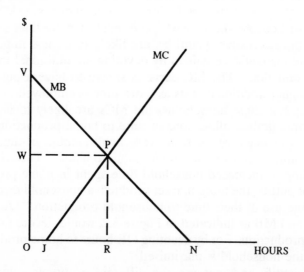

**Figure 3.2**          **Household equilibrium**

These different concepts of economic values are important for measuring the value of home production for the different purposes. Thus, if the purpose of valuation is toward national income accounting or Gross National Product (GNP), then the appropriate measurement is one of marginal valuation. This is because, conventionally, the value of each economic good for GNP purposes is taken to be equal to its market price or in other words, valuation at the margin. In the case of household production, this would be the value households place on the last hour performed in household production. Assuming no external effects, this value at the margin also reflects the true marginal opportunity cost in production. Consistent with GNP calculations, the product of the marginal value, and the total number of hours performed by the respective households, establishes the total GNP value of household production.

Thus, in prescribing an appropriate method of valuation of household production for social accounting purposes, the proposed method and its derived estimates should at least be shown to be consistent with this marginal valuation. In this respect, it appears that despite some of the problems mentioned earlier, the replacement cost method of using the market cost of paid general housekeepers to valuing unpaid self performed household services is the appropriate valuation per se. But there are still some adjustments which need to be made to these replacement value estimates (more on this in Chapter 5).

Of increasing practical importance is the valuation of household production for use in matrimonial property settlements. Very often, the division of matrimonial assets and properties are based on respective spousal contribution to the home. Such contributions are based on two sources -- (1) income and earning capacity and (2) household production. The contribution made by a spouse to the household in terms of household care and provision, may be interpreted as the foregoing of other activities -- earning income, pursuing leisure or other non-remunerative activities -- which he or she could have engaged in.

In this case, it can be argued that since time is required for household production, the relevant economic valuation of the time spent in housework can be derived from the values that are forgone by not putting the time to alternative uses. Thus, the time spent in household production has a cost -- an opportunity cost -- in terms of what has been given up. If it is taken that the spouse who provides these services incurs the opportunity cost, then the total opportunity costs of housework less his or her own consumption of the benefits produced, is the appropriate measure of the sacrifice made by this spouse to the household. Of course, this is not to say that the other spouse who undertakes market work and contributes to household monetary income has no opportunity cost. It is a weighting of their respective contributions and a measure of their respective 'sacrifices' in terms of opportunity costs less their own consumption of the benefits produced that would seem to yield the better estimate, and hence the more appropriate accounting for such purposes of matrimonial property settlements. Of course, there are other important considerations[54] which must be brought into such a calculation, but they are beyond the scope and theme of this study.

A third use in deriving estimates of the value of household production is in the area of welfare and compensation questions. Household production is usually seen to be of great importance to the welfare of households and even to be essential given that food needs preparation, children (if any) require care and some minimal level of cleanliness is necessary to maintain physical well-being. Household production is after all, the creation of goods and services by household members for their own benefit.

The notion of consumer's surplus is that people do value goods by some amount in excess of what they actually pay for them in market exchanges. This is to say that consumers would be willing to pay an amount in excess, and often far in excess, of the market price. The availability of goods at such prices allows consumers to enjoy a surplus from the exchange -- they obtain something by paying only a portion of what they would be willing to sacrifice for the infra-marginal units. This surplus amount is, to the consumer, a gain from the exchange and these gains are appropriately termed welfare measures in consumption.

Similarly, on the production side, the notion of producer's surplus is that producers are willing to supply goods if necessary, at prices below the market price for the infra-marginal units. But because all units are sold at the established market price, producers are said to gain from the exchange and these gains are appropriately termed welfare measures in production.

As a result of the peculiarities in the circumstances of household production, in that, households themselves produce and consume the

---

[54]   It might be argued that the opportunity cost of the spouse who spends most of his or her lifetime on household production falls over this time as it becomes harder to participate in the market due to a depreciation of skills. In such cases, the total opportunity costs incurred by the spouse would be seriously underestimated and an adjustment is clearly warranted to reflect the present value of the opportunities forgone at the time the decision to forgo them was made.

goods and services, the measure of household welfare from household production would be the sum of consumer surplus and producer surplus. This sum of economic surpluses is the net value measurement from household production.

This net value measure of household welfare, apart from establishing what the household gains from having household production over costs, also has another important use; this is in the area of tort litigation -- the settlement of legal disputes as to the amount of compensation required for losses in household production services due to wrongful injury or death which is clearly a welfare question.[55] What is involved, is the degree to which family welfare is affected as a result of the given tort. Thus, the appropriate measure for such compensation questions involving welfare loss through disability or death of a family member who provided household services is the amount required to restore the household's total net benefit position (or net value) to the level enjoyed prior to the accident. However, the literature on welfare and compensation questions most often uses either an opportunity cost measure or a replacement cost estimate; since both of these methods are not welfare measures, both would be inappropriate.[56]

We can sum up our discussion here by looking at Figure 3.3 below, Figure 3.3 is actually a reproduction of Figure 3.2 however, with some additions.

The value of the ORth hour placed by this household -- marginal value -- is $OW_1$ , while, the total value of OR hours devoted to household production and one which is consistent with GNP measurements is $OW_1.OR$ or $OW_1PR$. This amount $OW_1PR$, is the correct value of household production if the purpose of valuation is for national income accounting. The use of the simple and direct replacement cost methods may lead to an over or under estimation of the GNP value of household production depending on whether the choice of replacement wages is $W_2$

---

55 Economists and lawyers have both addressed this compensation question. But the methods adopted remained questionable and to a large part, the estimates are inflated -- a factor which may have accounted for their continued popular appeal. Both estimates of replacement cost and opportunity cost are not measures of welfare from a theoretical point of view and as such they should not be asserted as one. See for example, Komesar 1974, Lambert 1961, O'Connor and Miller 1972.

56 A detailed discussion on household production compensation and the law is found in my published paper in the *Osgoode Hall Law Journal* (1987).

or $W_3$. If choice of the replacement wage rate is $W_2$, the value of household production will be overestimated by the amount $W_2W_1PS$. Whereas, if $W_3$ is used, the value of household production will be underestimated by the amount $W_1W_3TP$. Strictly speaking, apriori, on an aggregate basis, we cannot say for certain whether which will dominate but it appears more likely that because not many households use paid domestic help services, the value of household production using the replacement cost method will be overestimated.[57]

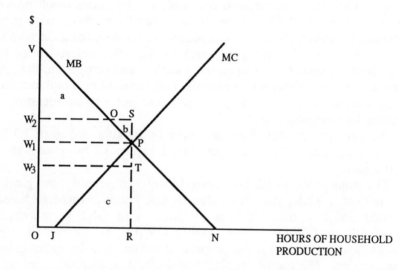

**Figure 3.3**        **An illustration of the argument**

---

[57] It might also be argued that if all homemakers offer their services to the market, the result would be a lowering of the market cost and hence the replacement wage rate of domestic help. An analogy here is one of valuing a large block of stocks at the prevailing market price. Because that price level only holds for an existing volume of supply and demand of stocks, releasing a large volume of stocks all at one time in the market would almost certainly depress its price. Therefore valuing a large block of stocks at its current market price would lead to an overestimation of the worth of the stocks.

The total opportunity costs in performing OR hours of household production is the amount JPR. JPR, measures the total sacrifices made by the spouse who normally provides the household production to the household and as such should be the correct value of household production for matrimonial property settlements.[58] A replacement cost estimate will however yield $OW_2SR$ as the value of household production (assuming $W_2$). This is clearly an overestimation by the amount $OW_2SPJ$. For reasons discussed earlier, there is moreover no conceptual basis for using the replacement cost approach if the objective is to provide a measure of spousal contribution to household production. The replacement cost approach is simply inappropriate.

Finally for welfare purposes, the correct measure of a family's welfare in household production is, as argued earlier, the net value measurement. In Figure 3.3, this is shown as the amount OVPJ. The replacement cost measure will however take $OW_2SR$ to be the welfare estimate; clearly an overestimation if a < b + c or an underestimation if a > b + c where a = $W_2VQ$, b = QSP and c = JPR. Suffice also to say that the opportunity cost measure is also inappropriate since it is not a welfare measure at all.

In sum, matrimonial disputes involving opportunities forgone involve different measures as do attempts to value household production for national income accounting purposes and for welfare and tort litigation. A clear specification of the purpose of valuation is clearly warranted. Unless this is so, the measurements obtained will be seriously in error.

Given the increasing importance being attached to estimates of the quantity and economic value of household production in policy formation, there would appear to be considerable point to appraisals that would yield more accurate and appropriate valuations. To this end, household production research is required to clarify and solve such immediate problems as the definition and quantification of household production, the treatment of joint-production activities, and the appropriate method of valuation. Without a serious attempt to at least specify and resolve these four problems, any household production study would undoubtedly be questionable and any estimates if generated, would remain estimates of curiosity.

The next chapter examines the major empirical works on household production measurement and valuation. While certainly not an exhaustive list of past studies, it aims to provide a reasonably comprehensive survey

---

[58]   Of course, across households and over time, opportunity costs change. Also see footnote 53.

of some of the more serious attempts to valuing household production and their estimates, and to examine them for some means of comparability and/or differences in methodology employed and in other results. The studies examined include those done in Asia.

# 4 Survey of previous major empirical studies on household production

With the exception of the United States, there are very few formal empirical studies on household production to be found for any country and none at all for Singapore. This chapter lists seventeen major empirical studies and a summary is provided for each of the study on the methodologies chosen and findings on the quantity and economic value of household production.

There are however, quite a number of informal empirical studies on household production yielding for the most part dubious estimates. These informal studies are what one commonly finds in magazines, newspapers and commissioned bank reports (Chase-Manhattan Bank, 1965; Ottawa Journal, 1966; The Globe and Mail, 1973; Changing Times, 1973; The National Enquirer, 1983; among others). Thus, among their improper and erroneous procedures used to measure and value household production include, carelessly worded definitions or no definition were given to what constitutes household production; ignoring the problem of simultaneous activities by double or even multiple counting of time; and using particularly high wages allocated to household functions for valuation purposes. According to these studies, the value of household production amounts to 60 to 70 percent of the GNP; clearly a very high figure. For these reasons and the fact that the objectives of these studies are more toward valuation for use in divorce and wrongful death litigation, we omit them from discussion here.

Also omitted are formal empirical studies which are very micro-oriented. Such studies normally involve taking small and/or select sample of households to test some hypothesis on household behaviour; trying out new methods of valuation of household production; and on measuring the worth of housewives other than for social accounting (Gronau, 1977 and 1980; Proulx, 1978; Brody, 1975; Morgan et al., 1966; among others).

All the studies reported here use either the replacement cost or opportunity cost method or some variant of both. Since these two methods were discussed in detail in the previous chapter, only the statistical-mathematical formula for each method and its variant are produced below.

## 4.1 Estimation formulae

The three estimation methods and their formulae discussed here are (1) the replacement cost by housekeeper method, (2) the replacement cost by specialised function method, and (3) the opportunity cost method. Also discussed are some of the variations of each of these methods and their estimation formulae.

*Replacement cost by housekeeper method, RCHK*

For this method, the assumption is that of hiring a single individual (maid or housekeeper) to do all the family housework. The simplest estimation methodology of this method is to use the annual average gross wages of housekeepers. That is,

$$H = \sum_{i=1}^{N} W^{RA} = W^{RA} \cdot N$$

(f1)

where H = value of home production in an economy; $W^{RA}$ = annual average gross wages of a hired housekeeper; and N = number of households in the country.

The advantage of using (f1) lies in its simplicity. There is no need to estimate the average number of hours a household spends on household production per unit time. All that is required for the estimation of the value of household production is knowledge of the annual wages of paid housekeepers and the number of households in the country.

A more difficult estimation formula of RCHK is to first estimate the average number of hours a household spends on housework per unit time (say per day or per week, etc.)[1]. And, second, estimate the average wage

---

[1]  This can be obtained through time use surveys or time-budget studies. A typical time use survey would pose questions to a member or members of a household
(continued...)

rate per hour of paid housekeepers. That is,

$$H = \sum_{i=1}^{N} (52 \sum_{j=1}^{n} T_j^{ph} W^R) \tag{f2}$$

where $T_j^{ph}$ = time spent in household production per week by the jth family member; $W^R$ = average wage rate per hour of a paid housekeeper; and n = member of household members in a family-household.

So that, the annual value of household production of an economy is estimated by taking the product of the total time spent in household production per week by all members of a household, and the hourly replacement wage rate of a hired housekeeper, aggregated over 52 weeks and over all households in the economy. Of course, the formulae (f1) and (f2) can be adjusted for disaggregation into many forms such as the variations in wage rates of hired housekeepers and the time spent on the basis of family size, number of children, location of household, and income differentials, etc.

*Replacement cost by specialised function, RCSF*

This method makes the assumption that a market replacement can be found for each type of housework. The time spent by household members in cooking, cleaning, washing and child care can be valued in terms of the wage rates commanded by people who offer such itemized services in the market. That is,

$$H = \sum_{i=1}^{N} (52 \sum_{j=1}^{n} \sum_{l=1}^{m} T_{jl}^{ph} W_l^R) \tag{f3}$$

---

[1](...continued)

asking for an estimate based on recall of the average total number of hours the family spends in each household chore on a daily or weekly basis. In contrast, a time-budget study is usually based on a family's diligent recording of actual time allocations in various household chores. These time recordings may last for a few days to a week. While the latter method of estimating time use is clearly preferred over the former as it provides relatively more reliable information on actual time use, its application is usually restricted to small size or target sampling due to the unavoidable nature of the degree of required involvement and cooperation of members of a household. Time-budget studies are also less useful in generating estimates of time use in less developed countries where literacy rates are low.

where $T_{jl}^{ph}$ = time spent in household production per week by the jth household member in housework item $l$; $W_{l}^{R}$ = simple or weighted average of hourly wage rates of the various market services equivalent to housework item $l$; and m = the number of housework items.

The average annual value of household production of an economy is then obtained through the summation of the product T and W for all housework items, household members and all households in the economy.

*Opportunity cost method, OC*

This method assumes that household members allocate their time in such a way as to receive the same value from their last hour in paid employment as they receive from their last hour in unpaid household production and that they are free to choose and vary the hours spend in market work and home work. In its simplest form, the formula for this method is,

$$H = \sum_{i=1}^{N} (52 \sum_{j=1}^{n} T_{j}^{ph} W_{j}^{m})$$

(f4)

where $W_{j}^{m}$ = the gross market wage rate per hour forgone by the jth household member.

The product between $T_{j}^{ph}$ and $W_{j}^{m}$ and summed over all individual household members and all households yields the value of household production in the economy. This method is also called the opportunity cost method before taxes or OCBT.

There are two variations of (f4) which take into account of personal income taxation on gross wages earned in the market and the necessary work-related expenses - costs of commuting, meals and other expenses - so as to yield the more realistic and appropriate comparisons between the net returns from time devoted to housework vis a vis market work.

Called the opportunity cost net of taxes method, OCNT, the formula for this first variation is,

$$H = \sum_{i=1}^{N} [52 \sum_{j=1}^{n} T_{j}^{ph} (W_{j}^{m} - t)]$$

(f5)

where t = the marginal income tax on an additional hour of work. That is,

$$t = W_{j}^{m} t_{r}$$

78

where $t_r$ = marginal tax rate. Thus, (f5) can be rewritten as

$$H = \sum_{i=1}^{N} [52 \sum_{j=1}^{n} T_j^{ph} (W_j^m - W_j^m \, t_r)]$$

or,

$$H = \sum_{i=1}^{N} [52 \sum_{j=1}^{n} T_j^{ph} W_j^m (1 - t_r)]$$

(f6)

The second variation is called the opportunity cost net of compensation method, OCNC and the formula is,

$$H = \sum_{i=1}^{N} [52 \sum_{j=1}^{n} T_j^{ph} \{W_j^m (1 - t_r) - C_j^m\}]$$

(f7)

where $C_j^m$ = the additional work-related costs incurred on an additional hour of work by the jth individual.

Strictly speaking, (f7) is the theoretically more appealing formulae for the opportunity cost method since in reality, in choosing between work at home and work in the market, people do take into consideration these work-related expenses.

With these formulae, section 4.2 describes in detail the objectives, methods of measurement and valuation, procedures, limitations and results of some of the major empirical studies done on household production. Although admittedly, the survey does not and cannot cover all studies to date -- due in part, to resource constraints and the inaccessibility of some non-English Language works -- it does include some studies which until now, are relatively unknown outside North America. These studies include those from Asia and Europe.

## 4.2    Some 'Formal' empirical studies on household production

The empirical studies discussed below are grouped in first, country studies and second, within each country study in chronological order of the year of publication. All money estimates are in US dollars unless otherwise stated.

*United States*

*(1) Wesley Mitchell, 1921*

This is the earliest known work on the valuation of household production. Titled, 'Income in the United States: It's Amount and Distribution, 1909-1919', the study's objective is to indicate some order of magnitude of the contribution of housewives to the national income. Using essentially RCHK or specifically, (f1), Mitchell took 'the average pay of persons engaged in domestic and personal service, a group that includes many other occupations besides female-domestics'[2] and multiplied this by the number of households with full time housewives in the nation.

In this study, there is no attempt to evaluate contributions made by other household members. Further it takes no account of the fact that households with hired domestic help were already included in the GNP and as such by including all households in the economy in the evaluation would result in an overestimation. Results of the study showed that the value of household production ranged between 25 and 31 percent of the 1919 United States's national income.[3]

*(2) Simon Kuznets, 1941*

Following the same procedures as Mitchell, Kuznets also uses RCHK and (f1) to estimate 'the magnitude of household activities within the domestic circle'.[4] Including only the home services of full-time housewives and thereby ignoring the contributions made by other members of the household, Kuznets reported that the value of housewives' services amounted to not less than 25 percent of the 1929 United States's GNP. Allowance was made for variations between farm and non-farm family households. The value of household production for farm family households was calculated by taking the average annual pay of farm workers (US$600) and for non-farm family households. Kuznets took the average annual pay of domestic servants (US$900).

---

2    Mitchell, 1921, pp. 59.

3    The study compared urban and rural households with lower average cost being attached to the latter.

4    Kuznets, 1941, pp. 432.

80

Both Mitchell's and Kuznets's study did not take into account of the part-time home services of working wives nor the home services provided by other family members. There is also no disaggregation into say variations between family households with different demographic and employment characteristics, thereby not much information can be obtained from the study. Admittedly, the two authors did qualify that their estimation was crude and that the estimates are merely indicative of the rough magnitude of the value of non-market household production.

*(3)    Ahmad Shamseddine, 1968*

This study attempts to measure the value of unpaid services of housewives in absolute and as a percentage of the United States GNP in the years 1950, 1960 and 1964. Shamseddine's method of evaluation is that of RCSF or specifically, (f3). Using two time surveys -- that of Walker, 1955 and General Electric, 1952 -- on hours devoted by housewives in six categories or items of housework[5], the value of household production was derived by taking the product of the time use of each housework item and the average wage rates (obtained from the U.S. Department of Labour) for female domestic workers corresponding to three broad categories of housework (general housekeeping, meal preparation, and child care). Further, the average time use for each housework item was adjusted for differences in family sizes, number and age of children, and housewife's employment status.

The results of the study showed that as a percentage of GNP, the value of household production amounted to 29.5, 27.3, and 24.1 percent for 1950, 1960 and 1964 respectively. This intertemporal comparison has suggested that the value of household production decreases with time since the magnitude appears to have dropped from 29.5 to 24.1 percent over a decade. Shamseddin has attributed this decline in the value of household production to three factors:    (1) the increase in paid employment participation of housewives in the economy, (2) the increasing commercialisation of housework, and (3) more home produced goods were purchased from the market. However, as not all household tasks were included in Shamseddin's study -- excluded are shopping for groceries, chauffeuring, among other things -- and the fact that a great deal of

---

[5]    The six items of housework used in Shamseddin's study are meal preparation, dishwashing regular care of the home, physical care of the children and other family members, washing and ironing.

interpolation and extrapolation were done on the available data, this decline in the value of household production from 1950 to 1964 must be treated as inconclusive.

## (4)   Ismail Sirageldin, 1969

Among the many objectives of this study are to estimate the value of non-market output, specifically, household work; the value of car services as part of family income; the value of labour time foregone because of sickness or unemployment; and the value of having more or fewer hours of market work. With respect to the valuation of household work, Sirageldin uses both RCSF and OCNT; first, separately and second, comparing their results. In using RCSF (or f3), the activities considered were regular household work, painting and repairs, sewing and mending, growing food, volunteer work, education, among other unpaid productive activities.

The novel ideas in Sirageldin's work are that in using OCNT, he not only took into account of adjusting hourly foregone earnings of individuals in non-market work for marginal tax rates[6] but also corrected these average hourly earnings for time taken in commuting to work[7] and for some non-equilibrium conditions such as people's dissatisfaction with institutional constraints on time allocation, unemployment and illness.[8] Further, for those housewives who lacked labour force participation, Sirageldin uses a multivariate analysis to predict and assign their hourly earnings.

The results show that if household production is valued by using the RCSF method, the estimated annual value is $3457 per average household whereas the estimated annual value is $3929 for the OCNT method. Of

---

6    Adjustment for marginal rate of taxation simply involves reducing the market hourly wages earned by the effective marginal tax rate for the different income earnings category. This is essentially given by (f5) or (f6).

7    Travel or commuting time to work is considered part of a person's total work hours so that hourly wages were derived by taking the person's total earnings and dividing it by total work hours inclusive of travel time.

8    To obtain information on disequilibrium adjustment, the study uses survey methodologies and involves questions such as 'how much more (less) would you like to work', and information on the status of employment of respondents and their health.

the latter method, this comes to about 50 percent of the family's disposable income. As a percentage of US GNP for 1964, the value of non-market home output is about 32 percent when using the OCNT method and about 28 percent for the RCSF method. It might be noted that there appears to be hardly any significant differences in the estimates using the two different methods of valuation.

### (5) William Nordhaus and James Tobin, 1972

In a much lauded effort, Nordhaus and Tobin embarked on an arduous task of constructing an experimental 'measure of economic welfare' (MEW) which allowed for corrections and/or adjustments to be made to our present national income and product accounts to take into account of welfare variables. Their adjustments to the GNP statistic include a reclassification of GNP expenditures as consumption, investment and intermediate; imputations for leisure, services of consumer durable and for the product of household work; and correction for pollution. Though admittedly crude and exploratory, the study serves to remind us of the shortcomings and of the inconsistencies of our national income accounting rules with respect to what is to be included and excluded.

While many of Nordhaus and Tobin's proposals on reclassification and adjustments to the GNP remained controversial, it is their recognition and attention given to non-market work activities that the authors deserve praise. On household production, the OCBT method corresponding to formula (f4) was used. Information on hours spent on housework was obtained from a 1954 study of time use cited in De Grazia. Estimates for MEW and its components were obtained for the years 1929, 1935, 1945, 1947, 1954, 1958 and 1965. The estimates on time use were assumed to hold for all these years. In calculating OCBT, the average hourly wages were disaggregated by sex, employment status, etc. and deflated by the consumer price index. The value of non-market household activity was estimated at $85,700 million or 42 percent of the 1929 GNP, and at $295,400 million or 48 percent of the 1965 GNP.[9]

---

[9] Note that in calculating the value of non-market activities, Nordhaus and Tobin have included the value of student time and the opportunity cost for the unemployed was set at zero. In a later paper by Hawrylyshyn (1976), the Nordhaus and Tobin's estimates were recalculated to exclude the value of student time and the value of non-market activities was re-estimated as $46,800 million for 1929 and $284,900 million for 1965.

## (6)   Kathryn Walker and William Gauger, 1973

Walker and Gauger are two Cornell University researchers who conducted extensive time budget studies in the Syracuse area, New York State, during the periods of 1967-1968 and later on in 1977. Amassing useful information on family-household production activities, the authors then attempted to generate the dollar value of household production in the United States using the RCSF [or (f3)] method.

As one of the limitations of the study, the sample of 1378 families consists of only two-person households made up of husband and wife and two-parent households with children. All other types of family-households were ignored, e.g. single-person households, non-family households and single-parent households, etc. The sample was disaggregated by employment status of the wife, the number of children under 18 years of age, and the age of the youngest child (or the age of the wife in childless families). It was found that in the case of unemployed wife households, the amount of time use in household production was about 65.7 hours per week and 45.4 hours per week for employed wife households. When presence or absence of children was taken into consideration, the amount of time use was estimated at 59.5 hours per week for those households with employed wives and no children, and 84 hours per week for those households with unemployed wives and with children. Time spent on household work by husbands only averaged 10.5 hours per week with some variation by family type.

In pricing the time spent in household production, average hourly wage rates were obtained for various household tasks performed by market workers and then applied to the amount of time spent by each family member. The results of the study showed that in the case of unemployed-wife households, the average annual value of housework was $7470 per household, and $5520 per household for employed-wife households.[10] The findings of the study also quite clearly show that it is the wife who contributes the larger share of housework both in time and in dollars, and that this was true whether or not she was employed. For the economy as a whole, the value of household production was estimated

---

[10]   In a later study by Walker and Gauger (1980), the authors found that the 1979 average annual value of housework in a family with two teenage children is $14,500 if the wife is unemployed, and $10,500 if the wife is employed. Except for higher average hourly wage rates (due to inflation), the same methodologies were used in valuation of household production by the authors as in their 1967-1968 study.

84

to be $204 billion or 26 percent of the 1967 United States GNP; and this excludes households other than two-person and two-parent with children households.

*(7)    Maurice Weinrobe, 1974*

Arguing that market-based GNP growth rates (exclusive of housework) have been overstated because of the increasing labour force participation of married women, Maurice Weinrobe of Michigan State University attempted to measure the value of household production of married women and the changes in this value over time. The period of study selected was 1960-1970 and the method of evaluation was essentially OCBT or (f4).

Time use data was obtained from the Walker and Gauger time budget studies and these include household production by full-time and part-time homemakers, and by employed wives. This time data is assumed to hold throughout the period of study.

In pricing the time spent on household production of full-time housewives, Weinrobe assigns an opportunity cost equal to the median full-time female earnings in the market. And, for those female employees who work full-time, the author assumes that they do no housework; clearly this is contrary even to casual observation and contradicts the Walker and Gauger time budget studies. The findings of the study revealed that non-market household production of married women is worth an annual average of 30 percent of the United States GNP. The fundamental argument in the study has been that because of a structural change in the relationship of market to non-market work of married women, the measured GNP statistics have been misleading -- the study found that the economic growth record has been overstated by some 0.20 percent to 0.25 percent. On the basis of real output per worker, this overestimation comes to about 10 percent.

However, there are a number of restrictive assumptions used in the study which when relaxed may alter the results significantly. One concerns Weinrobe's exclusion of all household members other than the housewife. It may be that although married women enter the labour force, some other household members help out in performing household tasks. To the extent that other household members take over some of the housework that were previously performed by the wife, the value of household production would decrease by less than that estimated by Weinrobe.

85

Related to one, is Weinrobe's implausible assumption that married women in full-time paid employment do no housework. Time budget studies have shown that household production continues to be done and not significantly decreased when the wife becomes employed.[11]

Last but not least is Weinrobe's assignment of median female earnings in the market to those full-time homemakers which may not be appropriate. If it is the case that full-time homemakers generally have lower skills and education than their full-time market counterpart, than the application of the median female earnings for those married women who do not work in the market would grossly overstate the value of their time at home.[12]

## (8)    Martin Murphy, 1978, 1982

In an effort to compare alternative methods of valuation, Martin Murphy derived estimates of the value of non-market household production using both the opportunity cost method and the replacement cost method. These estimates are then aggregated separately for the economy of 1960 and 1970. Time use data on household production was obtained from Walker and Gauger's time budget studies for two-person and two-parent with children households, and that of Alexander Szalai's multinational time budget study for single men and women.[13]

The time use data was further classified by the employment status of the wife, number of children under 18 years of age, and age of the youngest child. Household production was divided into five main tasks: (1) food preparation, (2) house upkeep, (3) clothing maintenance, (4) family care, and (5) others. To obtain one value of household production, the annual average hours spent on each of these five household tasks were multiplied by their respective market equivalent hourly wage rate. This constitutes the RCSF method or (f3). To obtain the other set of estimates of the value of household production, the average total number of hours allocated to all five household tasks were multiplied by the relevant

---

11    See Walker and Gauger, 1973 and 1980.

12    Should it be the case, however that married women who chose to remain at home did so because they value their time and hence of the benefits from household production more than that of their offered market wage rate, then, the median female earnings would underestimate the value of production for these women.

13    See Szalai, 1966

opportunity wage rate net of taxes. This is the OCNT method (f5) or (f6). The results of the study showed that if the RCSF method is used, the value of non-market household production is $185.3 billion or 36.8 percent of the GNP for 1960, and $335.6 billion or 34.3 percent of the GNP for 1970. On the other hand, if the OCNT method is used, then the value of non-market household production is $189.5 billion or 37.6 percent of the GNP for 1960, and $362.5 billion or 37.1 percent of the GNP for 1970. Murphy thus asserted that opportunity cost estimates of home production exceed replacement cost estimates by 1 to 3 percent of the GNP. Further, it was noticed that the ratio of household production to the GNP had declined slightly between 1960 and 1970.

The author concluded by arguing that if the objective of valuation of household production is for GNP accounting than the RCSF method is the appropriate method whereas if the objective of valuation is for welfare accounting then the OCNT method would be more relevant. This is because GNP is first and foremost an index of production and should be valued by market equivalent costs whereas welfare measurement involves utility considerations which are inherent in the OCNT method.

Later, in 1982, Murphy published another paper on household production; this time he uses five valuation methods to derive the aggregate and per person estimates of the value of household production in the United States for the year 1976.[14] The five valuation methods are RCHK [i.e. (f2)], RCSF [i.e. (f3)], and OC [using (f4), (f5) and (f7)]. His objective is then to compare these empirical estimates using these different methods of valuation.

Murphy's time data source is based on a 1975-1976 time use study by the Survey Research Centre of the University of Michigan.

The main findings of this 1982 paper by Murphy showed that first, the aggregate value of household production is large regardless of the method of valuation. In terms of percentage to GNP, the RCHK method yielded 31.6 percent (or $540 billion); RCSF, 44.1 percent (or $752.4 billion); OCBT, 59.5 percent (or $1015.4 billion); OCNT, 50.7 percent (or $865

---

[14] In the same year, Janice Peskin also published a paper on the empirical estimates of the value of household production in the United States for 1976 using the same procedures, time data and methods of valuation as Murphy. The results are thus identical to that of Murphy. However, Peskin added a section on variations in the value of household work across women and discussed how the value of household production is affected by employment status of women, number of income earners in a household, presence of children, age of women, and women's own earnings. See Peskin, 1982.

billion); and OCNC, 44.1 percent (or $751.8 billion). Second, that the aggregate estimates are highly sensitive to the choice of method of valuation. Thus, the difference between the highest estimate using the OCBT method and the lowest estimate using the OCNC method amounted to about $263 billion. Third, it was noted that the opportunity cost estimates generally yielded higher estimates than the replacement cost estimates. And, finally, the per person estimates tend to vary significantly by sex and the level of earnings for all five methods of valuation.

*(9)    John Kendrick, 1979*

The objective of Kendrick's paper is to present exploratory estimates for the major types of non-pecuniary economic activities which are currently omitted from the United States income and product accounts. Although the list of imputations is not as exhaustive as that of Nordhaus and Tobin's study,[15] it includes most of the major non-market activities. These are (1) unpaid household work, (2) other unpaid labour services such as the value of student time and volunteer work, (3) final products charged by business to current expense, and (4) rental values of non-business property. For our present purposes we are interested only in the values he presented for unpaid household work. Here, unpaid household work is defined and measured in terms of five major types of activity: food and beverage preparation (includes service and clean-up), care of home grounds and equipment, making and caring for clothing and home furnishings, care of family (includes transportation), and household management (such as record-keeping, shopping, etc). Time spent on household production by all household members, employment status and number of children are obtained from two time use surveys, namely that of Walker and Woods (1968), and Juster's University of Michigan time-use survey (also known as the Michigan data, 1975). The method of valuing housework chosen in Kendrick's study is the RCHK (f2) method. Unlike Kuznets which also used the RCHK method, Kendrick's estimates included every household member whereas in the case of Kuznet's study, only full-time housewives were considered. Total annual hours devoted to household production in all five activities were then multiplied by the average hourly compensation of household domestics.

Kendrick's estimates on the value of household production covered the

---

[15]    See Nordhaus and Tobin, 1972.

period from 1929 to 1973. Specifically, he presented estimates for the years 1929, 1948, 1966 and 1973. In percentage terms, the value of household production to GNP was 25.3 percent ($74.2 billion) for 1929; 27.7 percent ($98.1 billion) for 1948; 26.2 percent ($131.2 billion) for 1966; and 24.9 percent ($145.6 billion) for 1973.

*Canada*

*(10)    Oli Hawrylyshyn, 1978; Hans Adler and Hawrylyshyn, 1978*

First presented as a report for Statistics Canada on the viability of deriving dollar estimates for the value of household production in Canada for 1971, Hawrylyshyn published a second paper with Adler on the value of household production in Canada for 1961 and 1971, this time in the *Review of Income and Wealth.* Time use data were taken from the average of two Canadian Surveys -- Halifax and Toronto -- involving families with at least two parents.[16] Empirical estimates on the value of household work and on the relative contributions of females to males to household work were presented for both regional (provincial) and national aggregates. The population data were disaggregated on the basis of employment status of wives, the number of children in the family, and the age of the youngest child. Three alternative methods of valuing household work were presented; these being the RCHK [using (f2)] method, RCSF [or (f3)] method, and OCNT [i.e. (f5) or (f6)] method.

The studies' findings showed that the value of household production in Canada for 1961 amounted to $15,661 million (or 39.5 percent of the GNP) using the RCSF method; and $17,310 million (or 43.6 percent of the GNP) using the OCNT method. For 1971, the comparable values are $38,758 million (or 41.1 percent of the GNP) using the RCSF method; and $37,633 million (or 40 percent of the GNP) using the OCNT method. In the case of Hawrylyshyn's study, a separate estimate of the value of household production in Canada was made using the RCHK method. Using the RCHK method, the value of household work in Canada for 1971 was $31,935.28 million (or 34 percent of the GNP).

On the relative shares in household work by females and males, the authors reported that for 1961, females' contribution to household work

---

[16]    For the other types of family-households, Hawrylyshyn made the comment that 'some assumptions were made to estimate the values ... for these households' (Hawrylyshyn, pp. 22).

in Canada was 26.6 percent of the GNP using the RCSF method, and 29.1 percent using the OCNT method. As for males' contribution to household work in Canada for 1961, this was 12.9 percent of the GNP using the RCSF method, and 14.5 percent using the OCNT method. And in 1971, females' shares of housework were 27.7 percent and 27.2 percent of the GNP using the RCSF and OCNT methods respectively. For the males' share of housework in 1971, the values were 13.5 percent and 12.8 percent of the GNP using the RCSF and OCNT methods.

Clearly, the value of household production appeared to be relatively stable in Canada for the ten-year period. The authors contended that this was partly due to the low participation rate of married women in the labour market.[17] Also, except for RCHK method, the value of household production was not significantly different in the choice of estimating methodologies. In the case of RCSF and OCNT methods of valuation, the results seem to be not far from around 40 percent of the country's GNP whether in 1961 or 1971.

*United Kingdom*

*(11)   Colin Clark, 1958*

This is one of the earlier empirical works on valuation of household production in the United Kingdom. In this study, Clark took the average cost per resident of providing for and maintaining adults and children in state-run homes and institutions to approximate the value of household work per head of population. However, excluded from these average costs per resident are such things as cost of buildings, rent and capital charges, expenditure on food, clothing and other purchased goods. In the choice of adult institutions, Clark used only those institutions which housed between 30 to 50 adults, and in the case of children's institution, only those less than twelve residents were considered.

The results showed that the value of household production in the United Kingdom for 1956 was in the region of 30.8 percent of the GNP (or £7.01 billion). While the method of valuation comes closest to that of a replacement cost approach in that it essentially asks the question -- how much would it cost to replace the housework services normally provided at home? -- there are additional and more serious problems in this approach. One concerns the implicit assumption that the services

---

[17]   See Adler and Hawrylyshyn, 1978 pp. 338.

provided in the home and their costs of production are similar if not the same as those in an institution. Clearly, they are different where scale of economies are likely to occur in institutions than in private homes. Similarly, quality may also be different say in the case of caring for children in private homes than in public homes. On the other hand, inefficiencies normally associated with large institutions may affect the average costs of maintenance and thus provide an upward bias to the derived estimates. Thus, problems such as these must be taken into consideration in interpreting the estimates.

*(12)    National Council of Women of Great Britain, 1975*

Based on a 1972 report by the National Council of Women of Great Britain entitled, "Home Management and Family Living", the present work is an extension of that report which looked into the question of unpaid work undertaken by the homemaker and the value of this work in relation to the United Kingdom's GNP.

Time use data were based on a council survey on time spent in housework and disaggregated by number of children, hours spent on housework by each family member, and age of children. Also, a record of time spent in each of the varied number of household tasks was obtained. The method of valuation was essentially the RCSF method.

The study reported that for families with children, the value of unpaid homemaking in 1975 was £4732 per family-household, whereas for families without children, the value was £1934 per family-household. Summing the values per household for all households with and without children, the study reported that the total value of homemaking in the United Kingdom for 1975 was £56,324 million or 61.5 percent of that year's GNP.

*Finland*

*(13)    Ritta Santti et al., 1982*

This study is part of a wider research project carried out by the Research Department of the Finnish Ministry of Social Affairs and Health to determine the value of household production in Finland.[18] Time use

---

[18]    The study is actually titled, "Unpaid Homework: Time Use and Value" and represents part 8 of the larger research project called "House-work Study". See

(continued...)

data were based on a 1979 household survey by the Finnish Ministry involving 1525 households. A major part of the study was also to find out how much time was spent on unpaid housework in the aggregate, how much time was spent in the different housework categories, and what were the important factors behind those variations.

The study reported that the average time used for housework was around 7.2 hours a day or 50.4 hours a week per household with cooking taking up almost one-third of the total time spent on housework. The factors affecting time use in unpaid housework were reported to be (1) size of household, (2) day of the week (usually Saturdays), (3) number and age of children, (4) employment status, (5) socioeconomic status of the head of the household, (6) household income, and (7) type of municipality (more housework was done in rural areas).

In valuing unpaid housework, the author used the RCHK (f1) method. However, the average earnings of municipal home helpers were used rather than those of paid domestic help as the wage rate of market equivalents.The value of housework in Finland in 1980 was calculated to be 45,288 Finnish marks per household. Since there were 1,718,815 households in Finland in 1980, the total value of unpaid housework was 77,841.7 million Finnish marks. This represents 41.7 percent of the Finnish GDP in 1980.

*Belgium*

*(14)    G. Chaput-Auquier, 1959*[19]

The objective of this study was to attempt a gross evaluation of household work and to relate it to the gross national product estimates of Belgium. Time devoted to household work was obtained from two sources: (1) from a time use survey conducted by the Department of Applied Economics of

---

[18](...continued)

Official Statistics of Finland, Special Social Studies, Ministry of Social Affairs and Health, 1982. Also, see Suviranta and Heinonen, 1980 for a study of the value of unpaid home care of children under the age of seven in Finland in 1979; and Suviranta and Mynttinen, 1981 for the value of unpaid house-cleaning in Finland in 1980. The two studies reported that as a percentage of Finland's GNP, the values of home care of children and unpaid house-cleaning were 5.7 percent and 3.9 percent respectively.

[19]    This study was cited in Goldschmidt-Clermont, 1985, pp. 73.

92

Brussels' University , and (2) from a French survey conducted by Girard, 1958.

Selecting only households with three members and having standard home equipment, the average time spent in household work was reported to be between 42 to 66 hours per week. On the basis of this time information, the author first applied the standard RCHK method, and second, the Colin Clark's replacement cost by institutions method of evaluation. Chaput-Auquier reported that both methods yielded a homework valuation of BFR 131,000 to 206,000 million in Belgium in 1956. This amounted to approximately 25 to 30 percent of Belgium's GNP.

*Sweden*

*(15)   E. Lindahl et al., 1937*

A serious effort was made by the Swedish team of national income accountants to estimate the value of unpaid housework rendered by wives and daughters for possible inclusion in the national income estimates of Sweden. The method of valuation used was the same as that of Kuznets, namely the simple RCHK method [or (f1)].

Without a need to estimate time devoted to housework, the authors only required information on the number of women in Sweden who are over 15 years of age and who are not employed in the market, and the average annual wages of paid domestic servants. While the procedures used in valuation were rather crude and the estimates derived having only rough indications, the Swedish effort has at least initiated official recognition of the value of homemakers to Sweden and the implications of the distortion of the country's product and income accounts resulting from the omission of such non-market activities.

The authors reported that the value of unpaid household production amounted to 20 percent of Sweden's national income in 1930.

*Philippines*

*(16)   Robert Evenson et al., 1980*

In 1975, an interdisciplinary advisory group consisting of academics from the University of the Philippines at Diliman, University of the Philippines at Los Barios, Yale University and the University of North Carolina designed and implemented a household survey of 576 rural households in the Philippines. The survey attempted to obtain a wide range of household activities and behaviour in a single cross-section survey. The collection of a wide range of household data allowed a number of hypothesis testing relating to modern household economic theory (those of the New Home Economics) such as fertility behaviour, nutrition and work, time allocation, and in household production, among others. For our present purposes, we are interested in the methods and estimates the authors presented for household production in the Philippines[20].

Time use data were first divided into time spent in market production, home production and leisure. On household production, time was further disaggregated by farm and non-farm households. Variations in time use were explained by differences in the number and age of children, employment status of the wives; age and education of the spouses; wife and husbands' wage rate; among others. The authors reported that on average, the non-farm family spends 7.2 hours per day on homework (or 50.4 hours per week) and the farm household spends 5.34 hours per day on homework (or 37.45 hours per week). Thus, the non-farm households spent on average 1.86 hours more per day than farm households. The authors attributed this to the observation that farm households spend more time working on their farm (which activities are not classified as housework by the study's definition), and many of these households were reported to have paid secondary jobs.[21]

The method used to value household production was essentially the opportunity cost method. In the Philippines study, the average female market wage rate was assumed to represent a lower bound measure of the value of women's work in the home, and this wage rate was employed as the value of their time. For those full-time housewives, their potential market wage rate was used using predictive variables such as education and age.[22] A similar method was adopted in calculating the men's contribution to home production.

The results showed that while the average market income of families amounted to only 5,783 pesos per family per year in 1975, their average

---

20 For estimates involving other objectives, see Evenson et al. 1978 (nutrition and work); Popkin, 1978 (Breast-Feeding Behaviour); Rimando, 1976 (Health); Torres, 1976 (Household Technology); and Banskota and Evenson, 1978 (Household Economics in General).

21 Reported in Evenson et al. 1978, pp. 316-317.

22 Some aspects on methodology closely followed that of Gronau's (see Chapter 2).

home production was valued at 6016 pesos. In terms of Becker's full income, the value of home production per household in 1975 was 51 percent or about 2 percent higher than their market income.[23] The men's contribution to home production was estimated at 668 pesos, while the women's and children's contribution were 3287 and 2061 pesos respectively per household per year.[24] No attempt was made in this study to value household production in the Philippines in the aggregate.

*Japan*

*(17)  Economic council of Japan, 1973*

In 1973, the Economic Council of Japan officially attempted an exploratory study to construct an alternative system of accounts which will lead to an estimated measure of economic welfare in Japan. Titled 'Measuring Net National Welfare in Japan', the study has among its welfare estimates of various non-market activities, estimates on the value of household production in Japan. Only full-time housewives were included in the study; thus contributions to household production by other household members were excluded whether full-time or part-time. The study compared the value of household production for 1955 and 1970. The OCBT method was used to value home production; this involves the sum of the product of the average number of hours spent on housework (based on two household surveys, one in 1960 and the other in 1970) and the average hourly wage earnings of females in the market. The 'Survey of People's Living Hours' revealed that Japanese full-time housewives worked an average 48.7 hours per week in 1960 and 45.6 hours in 1970. The value of household production was evaluated at 11.2 percent of the GNP in 1955, and 8.7 percent in 1970.

---

[23]  Full income is defined as the sum of the value of market income and home income (imputed). In this case, the value of home production as a percentage of full income is 51 percent or 6016 pesos. For more discussion on the concept of full income, see Chapter 5.

[24]  In the case of children's contribution to home production, we excluded the study's imputed value of school time.

# Table 4.1
# Summary of previous major empirical studies on household production
# (Original and Unadjusted)

| Author, Affiliation, and Year of Publication | Country of Study | Time Period /Date of Study | Methodology | Per Household | | | Value of Household Production for all Households in the Economy (in Billions of US$, unless otherwise indicated) | Value of Household Production as a percentage of GNP (HP/GNP) |
|---|---|---|---|---|---|---|---|---|
| | | | | Ave. Hours per week | Ave. Hourly Value | Ave. Annual Value | | |
| Wesley Mitchell[1] National Bureau for Economic Research (1921) | United States | 1909-1919 | RCHK | --[2] | -- | $ 900 | $18.5 | 25-31[3] |
| Simon Kuznets, National Bureau for Economic Research (1941) | United States | 1929 | RCHK | -- | -- | $ 833 | $22.5 | 25 |
| Ahmad Shamseddin, Elizabethtown College (1968) | United States | 1950 1960 1964 | RCSF | -- | -- | -- | 1950 $84.1 <br> 1960 $137.7 <br> 1964 $153.3 | 1950 29.5 <br> 1960 27.3 <br> 1964 24.1 |

## Table 4.1 cont.

| Study | Country | Year | Method | | | | | | | |
|---|---|---|---|---|---|---|---|---|---|---|
| Ismail Sirageldin, University of Michigan (1969) | United States | 1964 | RCSF OCNT | 7-9,[4] 16-54[5] | -- | $3,457 $3,929 | RCSF OCNT | $177.7 $204.5 | RCSF OCNT | 28.0 32.0 |
| William Nordhaus and James Tobin Yale University (1972) | United States | 1929-1965 | OCBT | 46.9[6] 16.4[7] | $2.00[6] $6.08[7] | $4,870 | 1929 1965 | $85.7 $295.4 | 1929 1965 | 42.0 48.0 |
| Kathryn Walker and William Gauger, Cornell University (1973) | United States | 1967-1968 | RCSF | 65.7[6] 45.4[7] 59.5[8] 84.0[9] | $2.19[6] $2.34[7] | $7,470[6] $5,520[7] | | $204.0 | | 26.0 |
| Maurice Weinrobe Michigan State University (1974) | United States | 1960-1970 | OCBT | 65.7 45.4 59.5 84.0 | -- | -- | | $219.5 | 1960 1970 | 34.1 31.1 |
| Martin Murphy, University of Lowell (1978) | United States | 1960, 1970 | RCSF OCNT | 37.6 | -- | -- | RCSF OCNT 1960 $185.3 1970 $335.6 | RCSF OCNT $189.5 $362.5 | RCSF OCNT 1960 36.8 1970 34.3 | RCSF OCNT 37.6 37.1 |

**Table 4.1 cont.**

| | | | | | | | | |
|---|---|---|---|---|---|---|---|---|
| Martin Murphy, Bureau of Economic Analysis, Washington (1982) | United States | 1976 | RCSF RCHK OCNT OCBT OCNC | 48.9 | $3.83 | $9,748 | RCSF $752.4<br>RCHK $540.0<br>OCNT $865.0<br>OCBT $1,015.4<br>OCNC $751.8 | RCSF 44.1<br>RCHK 31.6<br>OCNT 50.7<br>OCBT 59.5<br>OCNC 44.1 |
| John Kendrick, George Washington University (1979) | United States | 1929<br>1948<br>1966<br>1973 | RCHK | 54.64[7]<br>70.47[6] | -- | -- | 1929 $74.2<br>1948 $98.1<br>1966 $131.2<br>1973 $145.6 | 1929 25.3<br>1948 27.7<br>1966 26.2<br>1973 24.9 |
| Oli Hawrylyshyn, Queens University (1978) and Hans Adler Statistics Canada (1978) | Canada | 1961,<br>1971 | RCHK, RCSF, OCNT | 47-84 | C$2.60 -$3.40 | C$6000 |     1961 1971<br>RCHK -- $31.94<br>RCSF $15.67 $38.76<br>OCNT $17.31 $37.63<br><br>(in Canada dollars) |     1961 1971<br>RCHK -- 34.0<br>RCSF 39.5 41.1<br>OCNT 43.6 40.0 |
| Colin Clark, Oxford University (1958) | U.K. | 1956 | RCI[10] | 44.2 | £0.25 | £570 | £7.01 | 30.8 |

**Table 4.1 cont.**

| | | | | | | | | |
|---|---|---|---|---|---|---|---|---|
| National Council of Women in Great Britain (1975) | U.K. | 1975 | RSCF | -- | -- | £4,732[11] £1,934[12] | £56.32 | 61.5 |
| Ritta Santti et al, Finnish Ministry of Social Affairs and Health (1982) | Finland | 1980 | RCHK | 50.4 | -- | Fm45,288 | Fm77.84 | 41.7 |
| G. Chaput-Auquier Brussel's University (1959) | Belgium | 1956 | RCHK[13] | 42-66[14] | -- | -- | BFR 131-206 | 25-30 |
| E. Lindahl et. al., University of Stockholm (1973) | Sweden | 1930 | RCHK | -- | -- | Kr.880[15] | Kr.1.65[15] | 20.0 |
| Robert Evenson et al, Yale University (1980) | Philippines | 1975 | OCBT | 50.4[16] | -- | 6016pesos | -- | -- |

## Table 4.1 cont.

| Economic Council of Japan, (1973) | Japan | 1955 1970 | OCBT | 48.7[18] 45.6[19] | -- | -- | -- | 1955 1970 | 11.2 8.7 |
|---|---|---|---|---|---|---|---|---|---|

Notes:

1. Quoted in Goldschmidt-Clermont, 1982 and Hawrylyshyn, 1976
2. A dash for this and other studies means no information was available or that information was difficult to ascertain
3. This refers to 1919 percentage values
4. For men
5. For women, depending on the number of children and employment status
6. Unemployed-wife households
7. Employed-wife households
8. Employed-wife households without children
9. Employed-wife households with children
10. Replacement cost by Institutions approach. What it would cost to replace housework services and care of children in institutions (see this Chapter)
11. For families with children
12. For families without children

13. He also used the Replacement Cost by Institutions approach but presented only one set of estimates using the RCHK method.
14. Households with three members and having standard household equipment.
15. Quoted in Hawrylyshyn, 1976
16. Non-farm households
17. Farm-households
18. For 1960
19. For 1970

**4.3        Some comparisons of the empirical findings**

The above studies purport to reveal estimates on the value of household production (summarized in Table 4A). With the exception of studies by Mitchell, Kuznets, Clark, Santti, and Lindahl, all other studies relied on information regarding time use on family-household production before imputing a value on that time. Studies of the former used the simple RCHK (f1) method where knowledge of only the average annual earnings of domestic help and the number of households in the country is required.[25]  In imputing a value on time spent in household production, studies of the latter used the RCHK (f2) method, RCSF (f3) method and/or the opportunity cost methods (OCBT, OCNT, OCNC). The merits (or demerits) of each of these methods have already been discussed (see this chapter, Section 4.1, and Chapter 3) as such will not be repeated here.

As all the studies were undertaken in various years and in current dollars, it would make little sense to compare them on the basis of the annual value of household production.   Further, the studies involved different countries and thus on a micro-level, the market conditions of hired-help and other market equivalents to household production activities would be very much different so that even if a sophisticated adjustment of the empirical results was to be undertaken for some means of comparability, there would be many distortions.[26]  Moreover in some studies, information on certain variables are simply not available. The absence of information thus requires skilful extrapolations or worse arbitrary assumptions.

That adjustments to the original empirical results of studies are difficult, and the resulting comparisons tenuous, it has not however deterred some researchers from trying.  One major attempt was made by Hawrylyshyn

---

[25]   Clark's study using the replacement cost by institution method is of course an exception.

[26]   Such is the case if, for example, in some countries, say the Philippines, Malaysia, etc, where because of cultural norms, women in general are expected to be full-time homemakers and as such more will be produced at home than say their counterpart in the United States or Canada where such cultural norms are less imposing.  But if the market wages of females in general are lower in the former and higher in the latter, then it may result in a conclusion that the value of household production in the former countries is less than those of the latter! Clearly a distortion.

(1976) to compare some 9 studies on household production;[27] but in arriving at some means of comparability, the author had to make a number of simplifying and arbritary assumptions, intra and extrapolations on time use, rural-urban cost of hired-help, extending the money value of household production derived from the study's sample to a larger sample, and extending the men-women share of housework obtained from a study's base sample to larger samples, among others.

Notwithstanding the difficulties involved in establishing some means of comparability for the varied studies, there are a number of 'safe' comparisons which one can make.[28] One, concerns the size of the value of household production in relation to the country's GNP measured in percentage terms. For the United States and Canada, the value of household production ranged from a low of about 24 percent to a high of almost 60 percent of the GNP. In the case of Europe, the magnitudes ranged from about 20 percent to 61 percent of the GNP. In the case of Japan, the value was about 9 percent of the country's GNP. Thus, with the exception of Japan, the value of household production averaged 34 percent of a country's GNP which is a fairly sizeable amount.

But even then, this average of about one-third of a country's GNP can be misleading, particularly since it is derived from a set of fairly large variation of averages. Again, Japan excepted, the value of household production for the largest percentage, 61 percent is more than three times that of the lowest percentage, 20 percent. The United States estimates alone ranged from 24 to 60 percent. While that of Europe, the range was between 20 percent and 61 percent. Clearly the disparity in value magnitudes of household production is by no means insignificant!

Before we jump to the conclusion that such studies on household production are unreliable, conjectural, and dismiss them altogether, it must be pointed out that there are good reasons for such a wide disparity

---

27  The nine studies covered are Lindahl et al., Clark, Weinrobe, Walker and Gauger, Nordhaus and Tobin, Sirageldin, Morgan, Kuznets, and Mitchell. See Hawrylyshyn, 1976. This chapter has excluded the study by Morgan, et al. (1959) for the United States simply because the study evaluated only a limited type of housework, namely food production and home repairs.

28  By 'safe' is meant that the empirical results are not subject to changes due to changes in say, inflation rates.

in measures.[29] The first, and obvious reason is that not all studies take the whole household as the sampling unit. Thus, studies like Shamseddine, Weinrobe and Lindahl exclude the housework that is done by the husband and all members of the family-household.

This reduces the quantity of household production reported in these studies and with it the value of household production as well, ceteris paribus.

A second reason for the disparity in values lies in the range of wages being used as replacement costs for those studies using the replacement cost methods. Instead of using the minimum or average wage rates for the various estimation of the value of each household function (which represents more accurately the generalist housewife), in some studies, particularly high and wages usually earned by specialists were used. This then produces an upward bias to the results obtained. The studies concerned include the National Council of Women of Great Britain, and Weinrobe. The seemingly large disparity in value measures can also be explained by the differences in and/or the lack of attention given to the definition of household production in some studies. This was further compounded by those studies which allowed for time spent to be counted twice for any two activities performed simultaneously whereas in other studies only time spent on the major task was reported. So that for those studies which relied on time use to value household production will differ in their resulting value magnitudes simply because differences in the amount of time spent in household production was reported, ceteris paribus. With the exception of the studies by Hawrylyshyn, Walker and Gauger, Murphy, Kendrick, and Evenson, almost all the rest of the studies have not made explicit their methodologies concerning definition and method of dealing with simultaneous activities. Worse, some studies have combined different surveys on time use without giving much thought to the problems regarding methodological differences in definition and the accounting of time spent in household production. Studies by Shamseddin, and Chaput-Auquier are examples.

Differences in valuation methodologies add to the disparity. This is particularly true for the opportunity cost methods. The use of the OCNC method to value household production would surely result in lower

---

29  Some of the discussions here are taken from two of my papers, 'What is the Value of a Wife?' in *Singapore Business*, Vol. 10 February 1986, and 'Valuing Family Household Production: A Contingent Evaluation Approach' in *Applied Economics* (July, 1987).

estimates than say the OCNT method which in turn, would be lower than the estimates produced by the OCBT method. It must be pointed out that a household member(s) decision to work in the market and/or remaining at home should depend on what will be their net return from paid work and home production. Factors considered in the former would surely include taxes and fixed costs arising from costs such as commuting, meals and perhaps the intangible disciplines of a work place. In this respect, the correct opportunity cost method should be the OCNC method rather than the OCNT and the OCBT methods. Only one study used the OCNC method and that was the study by Murphy. However, for aggregation purposes, the application of the OCNC method is most difficult since it requires knowledge of people's fixed costs in market work. Thus, for all practical purposes, the OCNT method would suffice.

**Table 4.2**
**Value of household production as a percentage of GNP**
**By method of valuation**

| Author and Year of Study | Method of Valuation | | | | |
|---|---|---|---|---|---|
| | RCHK | RCSF | OCNT | OCBT | OCNC |
| Hawrylyshyn (Canada, 1971) | 34.0 | 41.1 | 40.0 | - | - |
| Hawrylyshyn and Adler (Canada, 1961) | - | 39.5 | 43.6 | - | - |
| Murphy: | | | | | |
| (U.S., 1960) | - | 36.8 | 37.6 | - | - |
| (U.S., 1970) | - | 34.3 | 37.1 | - | - |
| (U.S., 1976) | 31.6 | 44.1 | 50.7 | 59.6 | 44.1 |
| Sirageldin (1964) | - | 28.0 | 32.0 | - | - |
| Average | 32.8 | 37.3 | 40.2 | n.a.* | n.a.* |
| Variance | 1.44 | 26.88 | 31.56 | n.a.* | n.a.* |

* n.a. = not applicable

105

A more interesting comparison is to look at studies which specifically dealt with empirical differences in valuation methodologies. Five studies -- Hawrylyshyn (1978), Hawrylyshyn and Adler (1978), Murphy (1978 and 1982), and Sirageldin (1969) -- compared the different valuation methods using the same base sample and for the same years. While the first two studies concern Canada, the remaining three are for the United States. Table 4B shows these five studies, their resulting values for household production as a percentage to GNP using the various methods of valuation.

In terms of averages, it appears that the value of household production is highest when calculated by the OCNT method followed by the RCSF method and lowest by the RCHK method. When examined even closer, taking each study by itself, the same conclusion holds that the descending order of the magnitudes of the value of household production produced by the methods, is that of OCNT, RCSF, and RCHK estimates, respectively. On average, the difference between the empirical value estimates of the OCNT method and the RCSF method ranges from about 1 per cent to 6 per cent, whereas between OCNT and RCHK, the difference is even larger, from about 6 per cent to 11 per cent. It appears that the latter's larger difference is also reflected in the empirical estimates produced by the RCHK and RCSF methods (about 7 per cent to 14 per cent).

While there might be many reasons which could account for this relative order of magnitude, two highly suggestive reasons are (1) in adopting the OCNT method, many researchers use the average or median female earnings as an estimate of the potential wage rate of full-time homemakers. Studies and even casual observations have showed that women who are in the labour force tend to be more educated and possess more work experience than full-time homemakers.[30]   Thus, the opportunity cost of not working for full-time homemakers is arguably lower than the median or average income of those women who are working in the market. This then produces an upward bias on the empirical results; and (2) it was argued earlier that the use of the RCSF method is likely to produce over estimates simply because specialist wages rather than generalist wages were used. The latter is more typical of the RCHK method.

On intertemporal changes in the value of household production, only the United States studies -- because of the number of studies involving

---

[30]   See Ferber, 1980.

different years -- permit some reasonable comparisons; but even then only a very rough approximation. Keeping in mind the differences in the methodologies used, Figures 4C, 4D and 4E present a plot of the value of household production as a percentage of GNP over time and by method of valuation.

Contrary to some widely held views that because of improvements in home technology -- electrical appliances, labour-saving equipment, etc. -- the amount of time devoted to household work should fall, and with it, the value of household production to total GNP since more time can now be devoted to the market as reflected in the increased labour force participation of women. This appears to be true for the period of 1960 to 1970 but not so after 1970 (see Figures 4C, 4D and 4E) no matter which valuation methodology was used. If an extrapolation of the empirical results were taken, then only Weinrobe's study would register a minor decline in the ratio of HP/GNP beyond 1970.

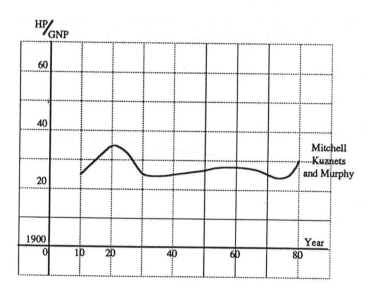

Figure 4.1    Value of household production as a percentage of GNP and by the RCHK method over time

107

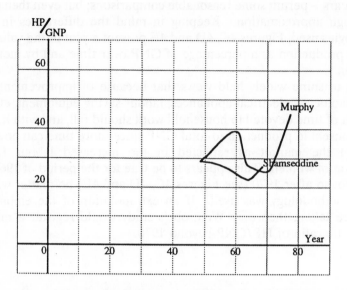

**Figure 4.2    Value of household production as a percentage of GNP and by the RCSF method over time**

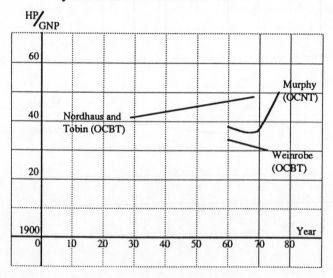

**Figure 4.3    Value of household production as a percentage of GNP by the OCBT and OCNT methods over time**

108

This assertion that HP/GNP seems to rise beyond 1970 is however very tentative and inconclusive since only a few data points were plotted. The question of whether time spent on housework has declined in recent years can be answered by examining studies on time use. In one such study (Walker, 1969), it was shown that the amount of time savings that has been achieved over the last few decades due to improvements in home technology has been very slight. Table 4F below reproduces the findings of Walker's study.

**Table 4.3**
**Time used for household work by urban homemakers in USA**
**(in hours, average of all days of the week)**

| Types of Activities[2] | Full-time Homemakers[1] | | | Employed Homemakers[1] | |
|---|---|---|---|---|---|
| | 1926-27 | 1952 | 1967-68 | 1952 | 1967-68 |
| All work connected with preparing and serving food | 2.8 | 2.6 | 2.3 | 1.9 | 1.6 |
| Care of the house | 1.3 | 1.6 | 1.6 | 0.8 | 1.2 |
| Care of clothes and laundry | 1.6 | 1.6 | 1.3 | 0.8 | 0.9 |
| Shopping and record keeping | 0.4 | 0.5 | 1.0 | 0.3 | 0.8 |
| Total | 6.1 | 6.3 | 6.2 | 3.8 | 4.5 |

Notes:     1.     Here 'homemaker' refers to housewives.
2.     Child care is excluded.

Source:     Kathryn Walker, 'Homemaking Still Takes Time', *Journal of Home Economics*, Vol. 61, no. 8, 1969.

In another study on time use by Newland, 1975, it was reported that there has been a very minor decline in housework time -- an average of 22 minutes per day -- from 1965-1975. That this was so, has been attributed by the author to demographic factors (such as employment status, number of children in the family, etc.) rather than technological factors (such as labour-saving equipment).

However in a study by Vanek, 1974 on time spent in housework in the United States, it was reported that:

> In 1924 such women spent about 52 hours per week in housework. The figure differs little (and in an unexpected direction) from the 55 hours per week for unemployed women in the 1960's. It is remarkable that the amount of time devoted to household work by such women has been so stable ...[31]

Thus, while Vanek's study seems to show a minor increase in the hours spent in housework vis a vis the other two studies (Walker, and Newland) which reported a minor decline in housework time, it appears that the differences are very small. So that, the only reasonable answer to the question of whether housework time has declined over the decades is that it has remained relatively stable.

This stability in housework time may be explained by two opposing factors. First, the reported increase in women's participation in the labour force has resulted in a fall in the total hours spent on housework.[32] This is evidenced by looking at Table 4A. In Walker and Gauger's study for example, the authors reported that unemployed-wife households spent on average, 65.7 hours per week on housework whilst for employed-wife households, this was 45.4 hours per week.

The second and offsetting factor to this decline in housework time due to increases in market work time is the notion that with the advent of labour-saving equipment and improved quality of life, people are now demanding more elaborate meals, more bought clothes and hence more washing, better child care, and a cleaner home environment; so that more rather than less time are devoted to housework (Walker, 1969; Vanek, 1974; Quah and Knetsch, 1982). There has also been some suggestions that for those employed-wife households it appears that much of the housework previously done on weekdays (when the wife was unemployed) is now reallocated on weekends (Ferge, 1974; Vanek, 1974).

In sum, the analysis on time trends and in the ratio HP/GNP remains

---

[31]  Vanek, 1974, pp. 116.

[32]  See Vanek, 1974; Weinrobe, 1974; and Walker, 1969.

indeterminate. While a cursory look at the plots given by Figures 4C, 4D and 4E seems to show an upward trend of HP/GNP beyond 1970, this must be interpreted with caution since intertemporal time budget studies have indicated a fairly stable use in the amount of time devoted to housework. Further studies are required involving more stringent criteria on adopting the same methodology in the definition and valuation of household production, standardizing wage rates of different years to a common base year, and using the same sampling unit -- in this case, the whole household rather than just the housewife -- for analysis, among other criteria before a study on intertemporal household production values and its magnitude in relation to GNP can be undertaken and its results considered more reliable.[33]

The only other non-United States study which looked at the ratio of HP/GNP over time is that of Hawrylyshyn and Adler, 1978 for Canada. Here, the authors have adopted almost all the stringent criteria mentioned above. Their results showed that the value of household production as a percentage of GNP rose from 39.5 percent in 1961 to 41.1 percent in 1971 using the RCSF method whilst the ratio fell from 43.6 percent to 40 percent using the OCNT method. The authors conclude that:

> No clear cut conclusions on the time trend of the ratio HW/GNP (value of household production to GNP) emerge from this paper, since, to some degree, whether that ratio increased or decreased depends on the valuation one chooses for HW. ... it is important to isolate the effects of changes in participation rates of married females, for they may be counterbalanced by other effects. On the whole, the ratio HW/GNP is fairly stable over time, and at all times forms a significant portion of GNP.[34]

On the principal factors which account for variations in not only the amount of household production but also its value, all studies showed the significance of such variables as the employment status of the wife, the number of children in a family, and the age of the youngest child. Thus, any future study on household production should at least include disaggregations of the data according to such variables.

---

[33] The study by Hawrylyshyn, 1976 is one such efforts, although as mentioned earlier in the text, there were problems in methodology and differences in sampling unit.

[34] Hawrylyshyn and Adler, 1978, pp. 348.

Finally, on the two rare Asian studies on household production, we see that the value of household production as a percentage of GNP are rather on the low side. In the case of Japan, the ratio HP/GNP was also declining with 11.2 per cent in 1985 to only about 8.7 per cent in 1970. However, it must be noted that in the Japanese study, only full-time housewives were included and all other members of the household as well as part-time housewives were excluded. This, then has the effect of underestimating the total amount of time spent in household production and with it, the value estimates.

Since the most recent study on the valuation of household production in Japan is that of 1970, it can be considered as relatively old. We have no information as to what has happened to the ratio, HP/GNP beyond 1970. However, one deduction has come from an independent survey conducted by the Japanese Government in 1985. The survey concluded that longstanding traditions of male-female roles in society will likely remain entrenched in Japan for many years to come; that this will be the likely trend is based on their findings that over 60 percent of men and 50 percent of women supported the notion that men should work in the market and women should remain at home. Another independent survey (1986) has come from an American insurance company (AIU) in Japan which revealed that among other things, a typical Japanese housewife spends about 50 hours at home attending to household chores, and based on the standard wage rate of domestic servants in Japan (i.e. the RCHK method using (f2)), the annual value of a Japanese housewife is about $14,220.[35] If these two surveys can be taken as anywhere near accurate, then they would suggest that the average time spent in housework in Japan has not changed much since the earlier study conducted in 1973 (see Table 4A).

The chapter has provided a reasonably comprehensive survey of the empirical literature on household production. In all the studies reviewed here, it appears that the value of household production as a percentage of conventionally measured GNP is large. The study of household production is thus important requiring further investigation. As most of the studies have originated from the West, very little is known on household production activities in Asia. The present study provides a useful starting point. The next chapter presents a model of household production leading to its valuation for use in social accounting.

---

[35] The Straits Times, June 20, 1986.

# 5 Modelling and valuing household production

In this chapter, a model for the valuation of market replaceable household production is first presented. Time is used as a measure of household production from which a money value of this time is imputed. The model indicates some of the limitations of the conventional approaches -- the requirement of restrictive assumptions concerning the nature of market and homework -- and establishes a theoretical basis of valuation from which imputed values for national income accounting purposes can be derived. The results from the model show that while using the market replacement wage rate may serve as a useful approximation of the value of market replaceable home production, a more accurate imputation would depend on whether the household is as efficient as the hired market replacement.[1] This variable, called the relative efficiency coefficient in the model, also has implications for households' decision to hire or not hire domestic help. The section concludes by showing how this relative efficiency coefficient can be estimated using survey techniques and contingent evaluation.

Recall from Chapter 1, that the total (complete) value of household production includes not only market replaceable household production (MHP) but also near-market replaceable household production (NMHP). The latter excludes much of the work usually associated with paid housekeeping but includes such things as commonly performed in households as home education of children, and family organisation and supervision in the home. In most studies on valuing household production for social accounting purposes, this relatively important aspect of family

---

[1]    While this point is not new -- it has appeared quite frequently as one of the major criticisms in using the replacement cost method to valuing household production (see for example, Gronau, 1977 and 1980; Hefferan, 1982 and Murphy, 1982) -- here, however it is shown formally in a general model.

life is often ignored and/or relegated to footnotes. While the estimation of NMHP is not easy and can be very problematic in view of the difficulty of identifying appropriate market equivalents, nevertheless some fairly crude estimation methodologies are presented in part two of this chapter (section 5.6).

But first, a look at market replaceable household production where a useful starting point of discussion is on prior models of household production; most notably those which centre around the concept of the household as a producing unit or otherwise, more commonly known as models of the 'New Home Economics.'

### 5.1    The 'new home economics': The household production model

Traditional microeconomics looks at the household as maximising a utility function consisting of goods and services subject to an income constraint. Recent developments have however shifted the view of the household from that of a consumptive unit to that of a productive unit in society. The 'New Home Economics' as it is sometimes referred to treats the household as a firm engaged in the home production of utility-yielding commodities, using market purchased goods and the time of family members as factor inputs (Becker, 1965; Michael and Becker, 1973). Thus, one major feature of the 'New Home Economics' is the importance given to the role of time in household decision-making models.

Figure 5.1 below illustrates the main differences between the traditional microeconomics of the household and the 'New Home Economics.'

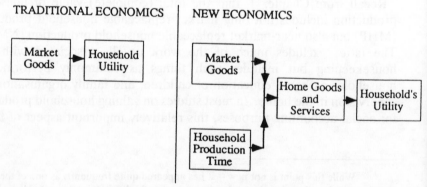

TRADITIONAL ECONOMICS | NEW ECONOMICS

**Figure 5.1** **Traditional microeconomics vs new home economics**

114

Thus, while traditional microeconomic theory views market purchased goods as directly contributing to household's utility, the 'New Home Economics' treats market goods as merely inputs to be combined with the time inputs of family members resulting in the final output of home goods and services. It is these home goods and services which then contribute to household utility.

Perhaps the best-known study which illuminated the ideas of the 'New Home Economics' came from the 1965 paper, *A Theory of the Allocation of Time* by Gary Becker and published in the *Economic Journal*. The point made by Becker was that economists have not given sufficient attention to non-market activities and the time these activities consume. On housework activities, Becker has, for example, incorporated time as an input and as a constraint into the traditional model.

Becker's model[2] takes the form of,

Max $\qquad U = u(Z_1, Z_2, ..., Z_n)$ $\qquad\qquad$ (1)

where $\qquad Z_i = Z_i(X_i, t_i, E)$ $\qquad\qquad\qquad$ (2)

subject to $\qquad\qquad T = t_w + \sum_{i=1}^{n} t_i$ $\qquad\qquad$ (3)

and $\qquad\qquad\qquad I = \sum_{i=1}^{n} P_i X_i$ $\qquad\qquad$ (4)

where $\qquad Z_i \qquad =\qquad$ the services or quantity of household 'basic commodity' i,

$\qquad\qquad t_w \qquad =\qquad$ time spent in the labour market producing market goods,

$\qquad\qquad t_i \qquad =\qquad$ time spent in producing household 'basic commodity' i,

$\qquad P_i$ and $X_i \qquad =\qquad$ the price and quantity of the market purchased goods used in producing household commodity $Z_i$

---

2 $\quad$ Except for some minor alterations, the household production model presented here is essentially drawn from Becker (1965), and Michael and Becker (1973).

$$I \quad = \quad \text{the household's money income, and}$$

$$E \quad = \quad \text{the existing technology in production}$$

As it turns out (3) and (4) can be combined into a single resource constraint representing the household's full income. That is, letting S to be the full income constraint of the household where,

$$S = WT + V = \sum_{i=1}^{n} (Wt_i + P_i X_i) + V \tag{5}$$

and $\quad W \quad = \quad$ wage rate from market employment

$\quad V \quad = \quad$ non-wage income

since $\qquad Wt_w = \sum_{i=1}^{n} P_i X_i$

The household's full income constraint (5) gives us the maximum income that could be earned by devoting all possible time (including consumption and home production time) to paid work.[3] The household's problem is then to maximise their utility function (1) subject to their full income constraint (5).

At the margin, each household commodity has a shadow price and in Becker's formulation, the shadow price of household commodity $Z_i$ is given by its marginal cost in production. The shadow price of $Z_i$ is derived from a weighted average of its goods intensity (dX/dZ) multiplied by the price of market goods (P) and its time intensity ($dt/dZ_i$) multiplied by the market wage rate of household members (W). Thus, the maximisation of the household's utility from household production yields standard results,

$$\frac{MU_i}{MU_j} = \frac{W \dfrac{dt_i}{dZ_i} + P \dfrac{dX_i}{dZ_i}}{W \dfrac{dt_j}{dZ_j} + P \dfrac{dX_j}{dZ_j}} = \frac{\Omega_i}{\Omega_j} \tag{6}$$

---

[3] This does not mean however working non-stop 24 hours a day as Becker do recognise that there will be some time allocated for eating and resting if money income is to be earned. See Becker (1965, pp. 94).

116

where $MU_i$ and $MU_j$ are the marginal utilities derived from commodities i and j respectively; $\Omega_i$ and $\Omega_j$ represent respectively the marginal costs incurred by producing i and j which are in turn determined by the shadow prices of market inputs (P) and time inputs (W).

These are familiar conditions of traditional demand and utility theory. The shadow prices are substituted for conventional market prices. The ratio of marginal utilities to shadow prices will be equated for all household commodities. Figure 5.2 below illustrates this using 2 household commodities, i and j.

It is also possible to derive the optimal use of amounts of inputs into the household production process such that maximum utility in terms of maximum output can be achieved. In equilibrium, this is shown as,

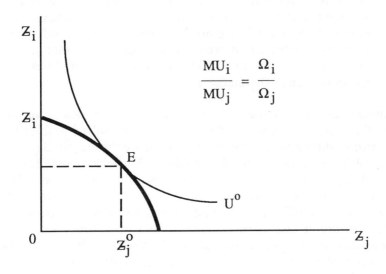

**Figure 5.2**          **Household equilibrium and household production**

117

$$\frac{\dfrac{\partial U}{\partial Z_i} \dfrac{\partial Z_i}{\partial f_{ik}}}{\dfrac{\partial U}{\partial Z_j} \dfrac{\partial Z_j}{\partial f_{jl}}} = \frac{MU_i \cdot MPP_{ik}}{MU_j \cdot MPP_{jl}} = \frac{P_{f_k}}{P_{f_l}}$$

where $MPP_{ik}$ and $MPP_{jl}$ are the marginal physical products of factors k and $l$ in the production of household commodities i and j respectively; $f_{ik}$ and $f_{jl}$ represent respectively the amounts of the factors k and $l$ used in producing i and j; $P_{f_k}$ and $P_{f_l}$ are the prices of the two factors k and $l$.

Following Michael and Becker (1973), a large number of testable empirical implications results from the household production model. Some of these are described by the authors as follows:

> ... households respond to the changes in the prices and productivities of factors, to changes in the relative shadow prices of commodities and to changes in their full real income as they attempt to minimise their costs of production and to maximise their utility. A reduction in the price of some factors of production will shift the production process toward techniques that are more intensive in the use of that factor and toward commodities that use the factor relatively intensely.[4]

Although Becker's household production model yields a variety of interesting results, there are, at the same time, unresolved problems. For one, extreme difficulties exist in trying to estimate the magnitudes of many of the critical variables in the model. Becker's 'basic household commodities' -- arguments in his utility function -- are not well-defined and the notion of these basic commodities can range from anything like cooked meals to religious accomplishments and sleeping; thus leaving much difficulty in distinguishing leisure from work activities. Further, there appears to be some confusion over the distinction between Becker's basic commodities and market goods which yield utility to the household. The criticism is that some goods are bought from the market for its own sake and do not combine with time at home to produce a basic commodity.

---

4   See Michael and Becker, 1973, pp. 139 for a more exhaustive list of empirical consequences which flow from the household production model.

118

Difficulties also exist in trying to estimate the shadow price of time for household members since in household production, what is involved is essentially non-market time. Using market wage rates of household members as proxies for non-market time valuation is unsatisfactory as will be shown later. Interestingly enough, contrary to Becker's emphasis on the role of time in household decision-making, he largely avoided the word 'activity' thus disregarding the possibility that work or a production activity which uses time inputs whether in the market or at home may actually contribute to the household's utility before a basic household commodity is created. An example of this might be the act of cooking a meal. The criticism is that cooking, child care and other productive activities may in themselves provide direct satisfaction or dissatisfaction.[5]

Thus, Becker's household production model assumes that time itself does not generate utility directly to the individual and hence in his model, there is no meaningful value of time. His value of time in household production is simply the opportunity cost of foregone market employment (i.e. the wage rate). In contrast to Becker's model, there are other models which treat time not only as a scarce resource but also as a heterogeneous final commodity in itself and which has different values in different uses. Thus, unlike Becker, the notion of different values of time such as week-day time, week-end time, etc., each having a different shadow price and on derivations of the value of time-savings is dealt in-depth by DeSerpa's 1971 paper, "A Theory of the Economics of Time."[6] Because of the complexity and perhaps even more inoperable empirically, Becker's theory remains preferred over that of DeSerpa's. Further, DeSerpa's model is more often used as the basis for most travel demand allocation of time models and as such need not concern us here.

It might be argued that theoretical papers should not be criticized for lacking in empirical applicability, but clearly most theories are aimed at specific application and as such more often than not, readers are left skeptical about how some of the variables can be estimated. But enough of carping criticisms. Although these and other difficulties to a large extent complicate empirical work, it cannot be denied that a major

---

[5]   For example, Pollak and Wachter (1975 pp 276) write: When the production of a commodity involves inputs of the household's time, the production process is likely to exhibit joint production. This is because the household derives utility or disutility from the time it devotes to each activity as well as from the nominal 'commodity output' of the activity.

[6]   DeSerpa, 1971.

contribution to economic theory has already been made. Of immediate significance of the 'New Home Economics' is that it clearly considers non-market activities within the home as contributing to utility, and hence having economic value just as those commodities -- goods and services -- purchased in the market. Becker's work is of considerable importance as it represents a pioneering attempt to formulate the household's maximisation problem, the emphasis on the role of time inputs and the problem of time allocation, all of which have given us a useful insight into the issues involved.

In contrast, the model below suggests that households derive utility from four activities:

1    consumption activities in the home (for example, drinking coffee and eating a meal at home, etc),

2    consumption activities in the market (for example, drinking coffee and eating in a restaurant, etc),

3    home production activities (for example, making coffee and cooking a meal at home, etc),

4    market production activities or market work (for example, working in a restaurant, commercial work, etc).

The time spent in household work by household members is used to measure household production. The rationale for using time to measure household production activities have already been covered in Chapter 3, Section 3.2. Suffice to say here that since for the most part household production activities consist of the provision of labour services which in turn requires time, household production can then be measured in terms of time. Besides, it is this aspect of the omission from the national income of the value-added of time spent in household production that is of our concern here.

Further, unlike existing formulations on household production which tend to concentrate on the dichotomy between market work and homework and the household's optimal allocation of time in these two sectors, the model below specifically includes the services of domestic help to substitute for some of the household production activities currently undertaken by household members. The model is formulated in such a way that work by hired-help gives utility by increasing household

120

consumption at home. Here, the model also allows for efficiency or productivity differences in household production between a paid domestic help who works only in the home and a household member who works both at home and in the market. Finally, an expression for the value of time spent in household production is derived from the equilibrium conditions of the model and is then used as a basis for the GNP imputation.

## 5.2 A market replaceable household production model

The family-household maximises a utility function,

$$U(R^{ch}, R^{cm}, R^{ph}, R^{pm}) \tag{1}$$

and
$$U_{R^{ch}} > 0, \ U_{R^{cm}} > 0, \ U_{R^{ph}} \gtrless 0, \ U_{R^{pm}} \gtrless 0$$

where (1) is assumed continuous, twice-differentiable and strictly quasi-concave. $R^{ch}$ and $R^{cm}$ are respectively, the home and market consumption activities of family members. For example, eating a home prepared meal would be $R^{ch}$ and eating a meal in a restaurant or buying a meal prepared by a restaurant and bringing it home for consumption would be $R^{cm}$. $R^{ph}$ and $R^{pm}$ represent home and market production activities of the household respectively. Cooking a meal and washing dishes at home are examples of $R^{ph}$ while working in an office and cooking in a restaurant are examples of $R^{pm}$.

Assume that the production of home consumption activities requires home goods ($Z^h$) such as cooked meals, made beds, etc.; the time family members spend on home consumption ($T^{ch}$); and market goods ($Z^m$) such that,

$$R^{ch} = R^{ch} (Z^h, T^{ch}, Z^m) \tag{2}$$

and
$$R^{ch}_{Z^h} > 0, \ R^{ch}_{T^{ch}} > 0, \ R^{ch}_{Z^m} > 0$$

and, $Z^h = Z^h (T^{ph}, Z^m)$ where $T^{ph}$ is the total time spent on household production and $Z^h_{T_f^{ph}} > 0$[7], $Z^h_{Z^m} > 0$.[8]

The production of home goods, $Z^h$, can be undertaken by both family members and non-family members such as hired domestic help or maid. The relative efficiency coefficient between family members and hired-help however differs and is assumed given by the following equation,

$$T^{ph} = k\, T_s^{ph} + T_f^{ph} \tag{3}$$

where k is the relative efficiency coefficient between family members and domestic help; $T_s^{ph}$ and $T_f^{ph}$ represent respectively, the time spent in household production by the domestic help and family members. Since $Z^h = Z\,(T^{ph}, Z^m) = Z^h\,(k\, T_s^{ph} + T_f^{ph}, Z^m)$, then assuming $Z^m = Z_o^m$ where $Z_o^m$ is some fixed amount of market goods input, the interpretation of k is as follows: when k > 1, implies that the portion of home output coming from domestic help is greater than that from family members given the same time input. For example, if $T_s^{ph} = T_f^{ph} = 1$ and k = 2, then $Z^h = Z^h\,(2(1) + 1, Z_o^m)$ where the time input of 1 from domestic help leads to a greater 'effective' output than the time input of 1 from family members. Domestic help in this case, is more efficient than family members.

Accordingly, cases of k < 1 and k = 1 imply respectively, that family members are more efficient than the hired domestic help, and family members are equally efficient as the hired-help. Equation (3) thus denotes the relative effective time inputs between family members and domestic help in producing home output and services. The higher the value of k, the greater the difference in household production efficiency between members and non-members of the household.

Assume further that home production activities of the family require market goods ($Z^m$) and the time spent in home production by family members ($T_f^{ph}$) such that,

$$R^{ph} = R^{ph}\,(Z^m, T_f^{ph}) \tag{4}$$

and

$$R^{ph}_{Z^m} > 0,\ R^{ph}_{T_f^{ph}} > 0$$

---

[7] At this point, I assume either linear-homogeneity or at least diminishing returns do not set in early.

[8] Same as note 7.

Finally, market consumption activities require market goods ($Z^m$) and the time family members spend on market consumption ($T^{cm}$); and market production activities require market goods ($Z^m$) and the time family members spend on market production ($T^{pm}$). That is,

$$R^{cm} = R^{cm} (Z^m, T^{cm}) \tag{5}$$

and
$$R_{Z^m}^{cm} > 0, R_{T^{cm}}^{cm} > 0$$

and,
$$R^{pm} = R^{pm} (Z^m, T^{pm}) \tag{6}$$

and
$$R_{Z^m}^{pm} > 0, R_{T^{pm}}^{pm} > 0$$

What (1) tells us is that, the family derives utility (disutility) from four sources of activities: the family's own home and market consumption, and home and market production. Production activities at home can be undertaken by both family members and through hired domestic help (subject to efficiency differences). Household production activities undertaken by family members yield direct and indirect utility to the household; the former in terms of enjoyment (disenjoyment) in the actual act of performing homework while the latter by increasing home consumption activities via increases in home output. Per contra, household production activities undertaken by paid domestic help only yield indirect utility in terms of increasing the home consumption activities of the household. This does not mean that the maid derives no pleasure (displeasure) from performing housework; all it means is that the direct utility (disutility) the maid derives from performing housework is of no concern to the household. Neither, does it mean that the maid does no consumption; all it implies is that the family just derives utility from the home production services contributed by the hired maid and not from the maid's consumption.

The household faces the budget constraint,

$$W^m T^{pm} = P^o Z^m + W^R T_s^{ph} \tag{7}$$

where $W^m$ = market work wage rate net of taxes

$W^R$ = exogenously determined market replacement

123

wage rate of domestic help, and

$P^o$ = exogenously determined market price of $Z^m$, where $Z^m$ is assumed to be a composite market good.[9]

Thus, (7) says that households' money wages (earned from market work), $W^m T^{pm}$, are spent on purchases of market goods, $P^o Z^m$, and payment of salary to domestic help, $W^R T_s^{ph}$.[10]

The total time constraint of the family-household is,

$$T = T^{pm} + T^{ph} + T^{cm} + T^{ch} \tag{8}$$

That is, household members allocate all the time available (eg. $T = 24$ hours per day) to market employment ($T^{pm}$), non-market household work ($T^{ph}$), and to home and market consumption ($T^{ch}$, $T^{cm}$).

The problem of the family-household is then to maximise (1) subject to the expenditure and time constraints given by (7) and (8). Forming the Lagrangean and substituting for equations (2) to (6), we obtain:

$$L = U \{R^{ch} (k T_s^{ph} + T_f^{ph}, T^{ch}, Z^m)^{11}, R^{cm} (Z^m, T^{cm}), R^{ph} (Z^m, T_f^{ph}), R^{pm} (Z^m, T^{pm})\} + \lambda_1 (W^m T^{pm} - P^o Z^m - W^R T_s^{ph}) + \lambda_2 (T - T^{pm} - T_f^{ph} - T^{cm} - T^{ch}) \tag{9}$$

The First Order Conditions[12] are:

---

9  The idea of composite goods came from Hicks (1936) and Leontief's (1936) composite commodity theorem, which asserts that if a group of market prices moves together in a parallel fashion, then the corresponding group of market goods can be treated as if it is a single good.

10  This assumes that the supply of domestic help (in time units) is perfectly elastic at the wage rate $W^R$. If follows, therefore, that $T^{ph}$ is the demand for domestic help.

11  Note that $Z^h = Z^h (T^{ph}, Z^m) = Z^h (kT^{ph}, Z^m)$ and $R^{ch} (Z^h, Z^m, T^{ch})$. So that, we can write $R^{ch} (k T_s^{ph} + T_f^{ph}, Z^m, T^{ch})$. Thus, there is no need for $R^{ch} [Z^h (k T_s^{ph} + T_f^{ph}, Z^m), T^{ch}, Z^m]$.

12  For simplification purposes, some chain partial derivatives in equations (10) to (17) have been reduced to simple partial derivatives.

124

$$L_{T_s^{ph}} = U_{R^{ch}} \, k - \lambda_1 \, W^R = 0 \quad [13] \tag{10}$$

$$L_{T_f^{ph}} = u_{R^{ch}} \, R_{T_f^{ph}}^{ch} + U_{R^{ph}} \, R_{T_f^{ph}}^{ph} - \lambda_2 = 0 \tag{11}$$

$$L_{T^{ch}} = U_{R^{ch}} \, R_{T^{ch}}^{ch} - \lambda_2 = 0 \tag{12}$$

$$L_{Z^m} = U_{R^{ch}} \, R_{Z^m}^{ch} + U_{R^{cm}} \, R_{Z^m}^{cm} + U_{R^{ph}} \, R_{Z^m}^{ph} + U_{R^{cm}} \, R_{Z^m}^{cm} - \lambda_1 P^0 = 0 \tag{13}$$

$$L_{T^{cm}} = U_{R^{cm}} \, R_{T^{cm}}^{cm} - \lambda_2 = 0 \tag{14}$$

$$L_{T^{pm}} = U_{R^{pm}} \, R_{T^{pm}}^{pm} + \lambda_1 \, W^m - \lambda_2 = 0 \tag{15}$$

$$L\lambda_1 = W^m \, T^{pm} - P^0 \, Z^m - W^R \, T_s^{ph} = 0 \tag{16}$$

$$L\lambda_2 = T - T^{pm} - T_f^{ph} - T^{cm} - T^{ch} = 0 \tag{17}$$

From the First-Order Conditions yield,

$$U_{R^{ch}} \, R_{T_f^{ph}}^{ch} + U_{R^{ph}} \, R_{T_f^{ph}}^{ph} = U_{T^{ch}} = U_{T^{cm}} = U_{T^{pm}} + \lambda_1 \, W^m = \lambda_2 \tag{18}$$

That is, at the margin the respective marginal utilities of home and market production, home and market consumption and the marginal utility of time are all equal.[14]

Further from (18),

$$\frac{U_{T^{pm}}}{\lambda_1} + W^m = \frac{\lambda_2}{\lambda_1} = V \tag{19}$$

where $\dfrac{\lambda_2}{\lambda_1}$ $(= V)$ is the marginal rate of substitution of time for money

and can be interpreted therefore as the price or value of time. Dividing (10) by k throughout (k > 0),

$$\tag{20}$$

$$U_{R^{ch}} - \frac{\lambda_1 \, W^R}{k} = 0$$

---

[13] $U_{R^{ch}} \, R_{T_s^{ph}}^{ch} = U_{R^{ch}} \, k$ since $R_{T_s^{ph}}^{ch} = k$

[14] Note that in equation (18), $U_{R^{ch}} \, R_{T_f^{ph}}^{ch} + U_{R^{ph}} \, R_{T_f^{ph}}^{ph}$ is not written as $U_{T_f^{ph}} + U_{T_f^{ph}}$ The reason is that by not simplifying the chain partial derivatives in this case, we clearly distinguished between utilities from consumption and production activities in the home resulting from time spent in home production by the family.

From (11),

$$U_{R^{ch}} + U_{T_f^{ph}} - \lambda_2 = 0 \tag{21}$$

where $R_{T_f^{ph}}^{ch} = 1$ and $U_{R^{ph}} R_{T_f^{ph}}^{ph} = U_{T_f^{ph}}$  [15]

Take (21) - (20)

$$U_{T_f^m} + \lambda_1 \frac{W^R}{k} = \lambda_2 \tag{22}$$

Dividing (22) throughout by $\lambda_1$,

$$\frac{U_{T_f^{ph}}}{\lambda_1} + \frac{W^R}{k} = \frac{\lambda_2}{\lambda_1} \tag{23}$$

Combining (19) and (23)

$$\frac{U_{T^{pm}}}{\lambda_1} + W^m = \frac{U_{T_f^{ph}}}{\lambda_1} + \frac{W^R}{k} = \frac{\lambda_2}{\lambda_1} = V \tag{24}$$

In equation (24), $\dfrac{T_f^{ph}}{\lambda_1}$ $(= MRS_{T_f^{ph}})$ is the marginal rate of substitution of

family home production time for money. Alternatively, it measures the money value of the marginal utility of time spent in household production by members of the household.[16] And $\dfrac{U_{T^{pm}}}{\lambda_1}$ $(= MRS_{T^{pm}})$ is the marginal

rate of substitution of time spent in market work for money. It measures the money value of time spent in market work.

---

[15] Keeping in that this reduction in the chain partial derivative comes from $R^{ph}$ and not from $R^{ch}$. It follows that $U_{T_f^{ph}}$ must refer to the utility from the act of performing housework using time inputs, $T_f^{ph}$.

[16] Recall also, that $U_{T_f^{ph}}$ in this case is derived from $U_{R^{ph}} R_{T_f^{ph}}^{ph}$ and as such, the $T_f^{ph}$ here refers to the time inputs and not to be confused with $U_{R^{ch}} R_{T_f^{ph}}^{ch}$ where the $T_f^{ph}$ of the latter measures home output in terms of time. See also note 15.

Equation (24) gives us an expression for the equilibrium value of time for the family-household measured in money terms. Specifically, it tells us that the value of time at the margin is equal to the market wage rate of family members plus the money value of time spent in market work (L.H.S. of (24)) and this is equal to the efficiency adjusted market replacement wage rate of a hired domestic worker plus the money value of the time family members spent in housework (R.H.S. of (24)).

Further, from equation (24), we can write,

$$\frac{U_{T^m}}{\lambda_1} + W^m - \frac{W^R}{k} = \frac{U_{T_f^h}}{\lambda_1} \tag{25}$$

Equation (25) has the interpretation that at the margin, the money value of working in the market and employing hired-help (L.H.S. of (25)) should equal the money value of working at home (R.H.S. of (25)).

### 5.3 Value of market-replaceable home production

The value of market-replaceable home production can now be imputed from the value of time spent in home production. The value of market-replaceable home production time, V is given by the right-hand side components of equation (24). That is,

$$V^1 = \frac{U_{T_f^h}}{\lambda_1} + \frac{1}{k} W^R \tag{26}$$

The value of market-replaceable home production time, $V^1$ is based on the value of that time at the margin. Thus, the present model would be appropriate for the measurement of the value of household production for national income accounting purposes. This is because conventionally, the value of each economic good or service included in the calculation of the GNP is taken to be equal to its market price or in other words, valuation at the margin. In the case of household production, this would be the value households place on the last unit of time devoted to household production. Consistent with GNP calculations, the product of this marginal value and the total number of hours (or units of time) performed by the respective households, establishes the total value of home production for an economy.

Further, in GNP accounting, not only is the value of the output measured at the margin, but also one does not measure whether the consumption or production of a particular service yields utility, disutility or zero utility. For example, in the production of a hair-cut in the market, no attempt is made to measure whether the service performed gives one utility or not. It is only the wage-rate received from the service provided or the price of the hair-cut that counts.

Thus, in equation (26) whatever assumptions made regarding $U_{T_f^s}$ should be ignored if the purpose of valuation of home production is for GNP accounting. Their inclusion is perhaps more suitable for measuring welfare and welfare changes at the margin.

By the above reasoning, the value of market-replaceable home production time at the margin reduces to,

$$V^1 = \frac{1}{k} W^R$$

The value $V^1$ now depends on what assumptions are made regarding k. If for example, k > 1 such that family members are less efficient than non-family members in home production, then

$$V^1 < W^R$$

So that using the market replacement wage rate would overstate the true value of home production time. On the other hand, if k < 1 such that family members are more efficient than non-family members in home production, then,

$$V^1 > W^R$$

The market replacement wage rate understates the true value of home production time. Finally, if k = 1, such that family members are equally as efficient in home production as non-family members, then,

$$V^1 = W^R$$

In this case, the market replacement wage rate becomes the true reflection of the value of market-replaceable home production time.

Assuming for the moment that we can empirically estimate what the average value of k is in a household survey (see Section 5.4 this chapter and Chapter 6 for hypothetical and actual derivation of k respectively), a reasonable estimate of the value of market-replaceable home production for national accounting purposes can then be given by,

$$MHP = \sum_{i=1}^{n} V_i^1 T_i^{ph} = \sum_{i=1}^{n} \frac{1}{k} W^R T_i^{ph} \qquad (27)$$

where MHP   =   value of market-replaceable household production in an economy

$T_i^{ph}$   =   the total time spent on market-replaceable home production by the ith household

$V_i^1$   =   the value of market-replaceable home production time as evaluated by the ith household, and

n   =   the number of households in the country.

Figure 5.3 reproduces Figure 3.2 and illustrates the derivation of the value of market replaceable household production for the ith household.

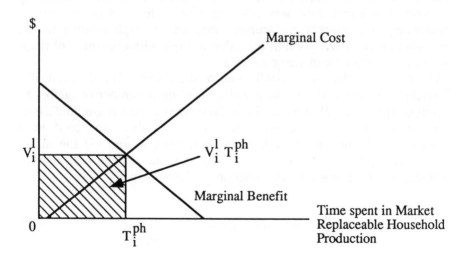

**Figure 5.3    Value of market-replaceable household production in household "i"**

**5.4**      **Implications of the present model on conventional measures**

One implication of the present model is that the conventional method of valuing the time spent in home production for national income accounting purposes using market replacement wages is only appropriate if one assumes that the household is as efficient as the market hired worker ie. k = 1. Clearly, this is a strong assumption and casual observations would indicate otherwise. The model thus confirms the often-heard criticism of the replacement cost method which uses the market wage rates of hired-help will lead to over or underestimation of the value of household production depending on how large and in what direction of bias are efficiency differences between the hired-help and the family-household.

Another commonly heard criticism of the replacement cost approach is that if households were actually required to pay the going wage rate for each hour of work, the amount of time allocated to household production would almost certainly be greatly reduced.

That not all of the home services get purchased implies that among other reasons, the household finds the price charged by replacements as being too high relative to their own opportunity cost of home production; that there exist significant efficiency differences in home productivity between household members and the hired-help; and/or the wages demanded by a market replacement may be too high relative to the household's own perception of the value of these services (some of these services may even be of marginal value).

The results of the model confirm all of the above. The decision of a household to hire domestic help will depend on a number of factors as equations (24) and (25) reveal. These factors are (1) the household's own opportunity cost in home production vis a vis the wage paid to a hired-help; (2) the relative efficiency coefficient, k; and (3) the utility and/or disutility associated with working at home and earning market income. In short, the demand function for hired-help, $T_s^{ph}$, is,

$$T_s^{ph} = T_s^{ph} (W^m - W^R, k, \frac{U_{T_f^{ph}}}{\lambda_1}, \frac{U_{T^{pm}}}{\lambda_1}) \tag{28}$$

Taking individually the factors given by (28), we would expect the household to hire domestic help if everything else remaining the same,

$$W^m - W^R > 0 \text{[17]}$$

$$k > 1$$

$$\frac{U_{T_{\hat{f}}^*}}{\lambda_1} < 0$$

$$\frac{U_{T_{f}^m}}{\lambda_1} > 0$$

But normally a household will not consider only one factor in its hiring decision. This can be seen from equation (25). The household will hire domestic help so long as the sum of the money value of working in the market and employing hired-help is greater than the money value of working at home. That is,

$$\frac{U_{T_{f}^m}}{\lambda_1} + W^m - W^R > \frac{U_{T_{\hat{f}}^*}}{\lambda_1}$$

The optimal number of hours demanded of a hired-help will be up to the point where

$$\frac{U_{T_{f}^m}}{\lambda_1} + W^m - \frac{W^R}{k} = \frac{U_{T_{\hat{f}}^*}}{\lambda_1}$$

Considerations involving all the factors in (28) will yield a welfare measure of household production. The value of household production minus the utility considerations on the other hand, gives us a GNP measure of home production. The hiring of domestic help by households and the wages paid for these services are clearly included in a country's national income accounts. However, what remains missing in these accounts are the corner-solution households and those households in which housework continues to be done even with the hiring of domestic help. Thus, it has been shown that a correct imputation for this value of market replaceable household production is the efficiency-adjusted replacement wage rate of a hired domestic help and not just the replacement wage rate. Using the conventional replacement wage rate of

---

[17] For an elaboration of this factor alone, see my paper in the *Singapore Economic Review, April 1986*.

domestic help without attempting to adjust for efficiency differences between the household and the hired-help will lead to serious biases.[18]

The other commonly used approach to value the time spent in market replaceable home production for GNP accounting purposes, namely the opportunity cost approach appears to be supported by the present model. That is, from equation (24),

$$V^2 = \frac{U_{T^m}}{\lambda_1} + W^m = V^1 = V$$

where $V^2$ = the value of time in market work. Again, for national income accounting purposes, the money value of $U_{T^m}$ is ignored and at the margin.

But this result will only be true if the valuation of home production time applies to homemakers who also engage in paid market employment. Significant problems arise when one considers individuals who work at home but who are currently not in paid employment. This applies to full-time homemakers who undertake the major if not all portion of home production so that over a long period of time, significant depreciation of market skills sets in. As such, it becomes not clear as to what the appropriate market wage rate is and has been foregone in pursuing household production.

One commonly used solution is to impute a potential wage rate for these full-time homemakers by taking the wage rate of women working in the market with similar characteristics (eg. education and/or skills). But there are serious problems with this method. One problem concerns the argument that since these full-time homemakers have chosen not to be employed, then they and their family must have valued implicitly their home production time much higher than the other households. A counter argument however is that these full-time homemakers have chosen not to be employed to avoid the usual disciplines of a workplace or equivalently, it is argued that for these people, they may use their time at home less

---

[18] Of course, we note that there are also quality differences between the output produced by the hired-help and that produced by the family-household. It is not for certain however that the quality of home goods and services produced by the hired-help is always less than those produced by family members. The quality could also be greater. See Chapter 3 Section 3.4.1, "Replacement Cost Methods". It is best to leave quality differentials alone since they relate more to subjectivity of the individual household and thus very difficult to make direct observations objectively.

intensively than in employment, again partly due to a loosen monitoring and intensive scheme. So while the method of using the wage rate to value home production time appears theoretically correct in the case of homemakers who are simultaneously involved in paid-employment, it would result in biased measure in the case of full-time homemakers.

Finally, one may argue that whatever the relative efficiency coefficient is, both the replacement cost and opportunity cost measures will lead to an overstatement of the value of time to low-opportunity (or less fortunate) households simply because many households are at corner-solutions and that they cannot afford to hire domestic help. This argument I contend is invalid for two reasons: (1) it suggests confusion in the objective of measurement in that the argument is only valid if one is measuring the implicit (welfare) value of time to the household. It is not valid when one is measuring household production defined as the provision of labour services and the creation of home output, and measured in terms of time units. Essentially it is the housework that counts and not the household's own welfare-valuation of time per se; and (2) for social accounting purposes, the valuation of a good or service does not require knowledge of whether households can or cannot afford them. It is the market value of the good or service and not the accessibility to these goods and services that is of relevance to the national income accountants.

It can be noted too that in the present model, this argument would not arise since there is already a budget or money constraint (equation (7)) which allows for this problem. While the argument may not be relevant when it comes to the valuation of household production for social accounting purposes, it is certainly of significance in the determination of household's decision to hire outside help. Given poor market employment opportunities and/or low opportunity cost of time, we would expect that for these households their probability in hiring domestic help would be very low. This can be seen from equations (25) and (28) of the model.

### 5.5    Estimating the relative efficiency coefficient, k

Since the model presented here is aimed at a specific application namely, to derive the value of market replaceable household production for national income accounting purposes using the efficiency adjusted market replacement wage rate of domestic help, then, the success in actually deriving an empirical estimate will depend on methods which can be used to evaluate the relative efficiency parameter, k.

One method is to utilise survey instruments and contingent questions. Consider two hypothetical scenarios:

*Case 1*
Total Replacement

Time Period 1 :    In this period, a household is assumed to do all its household chores and producing a given output, $Z_o$ using $T^{ph} = T_f^{ph} = 10$ hours a week. By this assumption, $T_s^{ph} = 0$ hours.

Time Period 2 :    Now the same household hires domestic help to engage in household production and does no household work for itself. Let us assume that the same output, $Z_o$ is produced but using $T_s^{ph} = 4$ hours a week.

The value of k is then,

$$k = \frac{T^{ph} - T_f^{ph}}{T_s^{ph}} = \frac{10-0}{4} = 2.50 \qquad \text{(see equation 3)}$$

*Case 2*
Partial Replacement

Time Period 1 :    Again, the household is assumed to do all its household chores and producing a given output, $Z_o$ using $T^{ph} = T_f^{ph} = 10$ hours a week. As before $T_s^{ph} = 0$.

Time Period 2 :    Household now hires domestic help but continues to do $T_f^{ph} = 2$ hours of housework a week. Let us assume that the hired domestic help does the remaining of what used to be an 8-hour output in 3 hours a week.

The value of k is then,

$$k = \frac{T^{ph} - T_f^{Ph}}{T_s^{ph}} = \frac{10-2}{3} = 2.67$$

134

Certainly with proper sampling procedures and interview methods, the above information on the values of $T^{ph}$, $T_f^{ph}$ and $T_s^{ph}$ can be obtained and a rough estimate on k generated. Of course, computing the value of household production requires not only the knowledge of the value of k but also the market replacement wage rate of domestic help. And, since both k and $W^R$ may vary quite significantly across the population (e.g. by size of household, number of children and location), a weighted average must be taken (this is especially true for large countries with scattered concentration of population).

### 5.6 Near-market replaceable household production: Home education and household management

These activities measured as 'secondary time for non-physical care of family members' in Walker's (1976) household production study, often go unnoticed but yet, constitute an important part of family-household function. Called 'invisible household production' by Paolucci (1977) of Michigan State University in a paper presented at a National Science Foundation Conference in Paris, these home activities include teaching children basic skills, helping children in their school work, household management, organisation and supervision of household tasks, among others. First, on home education.

*Home education*

A certain part of household production consists of providing simple home education to one's children. The end result of this activity is a form of human capital investment. In principle, this is not much different from Kendrick's (1979) measurement of investment in human capital for the United States by summing up the costs of providing formal education. The study by Leibowitz (1974) on home investments in children and how they affect the women's labour force participation gave recognition to this hitherto obscure role of the family-household. However, Leibowitz made no attempt to measure the value of such home education for social accounting purposes and it was Hawrylyshyn (1976) who attempted a measurement for Canada. Labelling this role of the family-household as 'tutorial child-care', Hawrylyshyn's study reported that it comprises more than 10 percent of measured total household production. Further, studies such as Walker (1976), Leibowitz (1973 and 1974), and Stafford and Hill

135

(1973) have reported that for higher educated-wife households, they tend to spend more time in child-care. That there is this difference in the time spent in child care between higher and lower educated-wife households can be attributed to perhaps the greater amount of time spent in home education of their children by the former over the latter.

Without attempting to measure the amount of benefits actually received from such investments in human capital at home -- attempts that would require human capital models and analysis of life-cycle variations in human capital investment, and not the focus of the present study -- it is recognised instead that there exists this role in the family-household similar to that of the role of teachers in kindergartens and primary schools.[19] An estimation of the economic value of such home education services provided by and for household members themselves would be to use the market-equivalent wage rate of such teachers. That is,

$$HE_i = T_i^e \, W^{RT} \qquad (29)$$

where $HE_i$ = Value of home education in household i

$T_i^e$ = Time devoted to home education in household i

and $W^{RT}$ = Exogenously determined wage rate of primary cum kindergarten school teachers.

No doubt a crude estimation methodology and one that involves no attempt to adjust for quality differentials between a school teacher and the family-member, it does however serve to remind us of this relatively ignored but important component of family-household production. Further, that no attempt is also made to adjust for efficiency differences between the market equivalent and that of family members (unlike market-replaceable household production) in providing home education services, is because of the extreme difficulty in establishing actual output in the form of human capital accumulation -- benefits which accrue only in the long run -- thereby no reference point is possible for use in efficiency comparisons between a market worker and a home worker.

---

[19] We exclude secondary and university teachers for the main reason that for most households the provision of home education is likely to be limited to small children (ie. less than 12 years old) although there are, of course, acknowledged exceptions.

We turn now to the other component of near-market replaceable household production, namely that of household management.

*Household management*

By household management is meant the organisation and supervision of household tasks including some matters of paper work such as preparation and keeping of financial records, grocery shopping lists, settling of bills and other household planning work. Thus, there are times when household production must meet deadlines and other time constraints -- say, children must be fed before they leave for school, domestic-help if hired must be supervised -- all of which involves a great deal of planning, coordination and organisation by at least a member of a family. Casual empiricism would confirm that household management is an important part of household production for it brings together some order from a wide variety of household production activities.

Indeed the household management role was observed as early as 1934 when Margaret Reid of the University of Chicago wrote in her classic book, *Economics of Household Production* that household production activities can be classified into two primary categories: Household Performance and Household Management. While household performance relates to physical activities in the home -- the cooking, cleaning, washing, etc. -- household management on the other hand is concerned with decision making, formulating policies and resource allocation.[20] It was Reid who defined household production in terms of whether such activities can be delegated to someone hired on the market and outside the household group. Reid argues that both household performance and household management can be replaced by market goods and services and as such constitute measurable household production.

However, this household management role has been relatively ignored over time while resources were directed toward the measurement and valuation of household production in terms of household performance. It was only until recently that recognition of the household management role became important and used as one of the main criticisms of the replacement cost method. Thus, Firebaugh and Deacon wrote,

A problem associated with the (replacement cost) method is that hired housekeepers usually do not carry out all the activities for the

---

[20]   Reid, 1934, pp. 11.

household and time and effort must be spent in supervising the person as well.[21]

and this was echoed in Zick and Bryant that,

> ... the market alternative cost approach excludes the management component of homework. Even if a household were to hire market substitutes, someone in the home would need to supervise their work and the value of this supervision has <u>not</u> been included in most market alternative calculations.[22]

The household management role has also became increasingly important in view of both structural (increased labour force participation of women, commercialisation of housework activities) and technological (introduction of labour-saving equipment for use in homework) changes such that more planning and coordination of household activities are now required. This has led Beutler and Owen to remark that : "With the increased number of goods and services to be coordinated, ... the homemaker's role is much more management oriented."[23]

While granted that the management role of households has become increasingly important in recent times, the next obvious question is how do we measure and value such an activity? As mentioned earlier, previous studies on household production have concentrated on measuring the physical aspects of household production and only minor verbal references are made to the household management role. Only in some studies are there attempts to measure this household management role in terms of time spent on matters related to paper work and shopping for groceries (Walker and Woods, 1976; Hawrylyshyn, 1978; among others). Following Walker and Woods (1976), we identify household management as comprising of household supervision of household tasks and paper work relating to household matters.

However unlike other studies where the activity 'shopping for groceries' is considered part of household management, in the present study it is considered as part of household performance. Shopping for groceries is

---

21  Firebaugh and Deacon, 1980, pp. 61.

22  Zick and Bryant, 1983, pp. 134.

23  Beutler and Owen, 1980, pp. 24.

a physical activity and follows what has been already planned in the form of a grocery list. It is the planning for such grocery lists that should count as part of household management and **not** when one is actually carrying out the activity of shopping for the items found on the list.

One very crude estimation methodology for valuing household management is simply to use the market equivalent wage rate of managers of small firms as the base of the valuation. It is the contention here that many household management activities in fact bear resemblance to that of activities normally performed by these managers. In addition to carrying our their paper work, other functions performed by these managers include settling some of the monthly office-related expenses and generally supervising junior clerks and maintenance personnel; all of which are similarly performed by household managers. Thus, the value of household management activities is given by,

$$HM_i = T_i^m W^{RM} \tag{30}$$

where $HM^i$ = the value of household management in household i

$T_i^m$ = the time spent in household management by household i

and $W^{RM}$ = the wage rate of managers of small firms

*In sum*

The value of near-market replaceable household production for household i, NMHP, can then be obtained by,

$$NMHP_i = HE_i + HM_i = T_i^e W^{RT} + T_i^m W^{RM} \tag{31}$$

While admittedly crude, at least by measuring (31) gives us a more complete picture of the kind of labour services one normally finds in households; whose activities have near market analogues and therefore capable of pecuniary imputation.

## 5.7     Value of household production

The total value of household production in an economy can now be obtained by summing up the values for MHP and NMHP. That is,

$$VHP = MHP + NMHP = \sum_{i=1}^{n} \frac{1}{k} W^R T_i^{ph} + \sum_{i=1}^{n} (T^c W^{RT} + T_i^m W^{RM}) \quad (32)$$

While recognised as an imperfect measure of the value of household production, it does however capture the main components of household production. The method proposed here differs from existing methods of valuing household production in that :

1   It uses both variations of the replacement cost method. That is, by housekeeper approach (MHP) and by specialised function approach (NMHP). It is argued here that the housework associated with MHP have easy identifiable market equivalents and can generally (and more realistically) be performed by hiring a single domestic help rather than an 'army' of specialised hired-help. On the other hand, the housework activities associated with NMHP are not normally included in the hired services of a single domestic help (nor are they appropriate for this kind of work). As such, it would be incorrect to impute a value for NMHP by taking the average wage rates of domestic help just as one does for valuing MHP. Here, the more appropriate market analogues are those of primary and kindergarten school teachers, and that of small firm managers and the value of NMHP can be imputed by taking the average wage-rates of these two market analogues.

2   The derived value of household production (equation 32) is question-specific in that the method is appropriate for social accounting purposes only. Thus, while the model presented for valuing MHP is formulated generally -- allowing for differences in utility consideration -- only the results of the model minus the utility considerations are used as the basis for the imputation of the GNP value of household production. The emphasis is on the housework performed and not on the implicit value of the act of working.

3    In measuring MHP, a relative efficiency coefficient (k) is introduced which accounts for differences in efficiency performance in housework between the hired-help and that of family members. Thus, it was shown that depending on the value of k, the value of household production using the conventional replacement cost method will be overestimated if k > 1 or underestimated if k < 1. A more accurate imputation would thus require an efficiency-adjusted replacement cost method.

However, just as there are advantages to using the present methodology in valuing household production for social accounting purposes, there are at the same time, some limitations. It will be recognised, that the model and method of estimation draw from both neoclassical economics and the 'new home economics.' Both paradigms are not without problems. Some of the major problems associated with the present approach include:

1    The strong assumption that there exists a 'joint' or one household utility function so that every household members' preferences are said to be included in that function. Thus, the use of the household as the unit of analysis obviates major conceptual problems in aggregating individual member's utility function.

2    A second and closely related assumption is that households maximise and as such are always making decisions at the margin. This implies that either households are quick to respond to a changing environment or that the environment itself is stable.

3    The assumption that the household must at least evidence constant returns to scale or else diminishing returns only at a much latter stage of production. If this assumption does not hold then, there would be significant problems in the use of time as a measure of household output.

Clearly these are serious problems which require attention if not at least acknowledgement. First, the postulation of a household utility function is theoretically correct since the relevant unit of analysis is at the household level. As Mahoney (1961) has pointed out:

We start with the concept of the spending unit defined as a household of persons who pools their resources to make joint

141

decisions concerning the expenditure of those resources .... The welfare of the entire spending unit is assumed to be the criterion for decision making.[24]

Besides, in accounting for the value of household production in an economy, surely every household member's effort and contribution ought to count. And, in defending the existence of a household utility function, Samuelson (1956) wrote:

> Where the family is concerned the phenomenon of altruism inevitably raises its head: if we can speak at all of the indifference curves of any one member, we must admit that his tastes and marginal rates of contribution are contaminated by the goods that other members consume. Those ... external consumption effects are the essence of family life. ... if within the family there can be assumed to take place an optimal reallocation of income so as to keep each member's dollar expenditure of equal ethical worth then there can be derived for the whole family a set of well-behaved indifference contours, relating to the totals of what it consumes; the family can be said to act as if it maximises such a group preference function.[25]

Further, Becker (1981) in responding to critics of the 'New Home Economics' has posited a head of the household whose own-well being depends on the well-being of all other family household members.[26] Thus, defining altruism in this manner, Becker argues that all household members will attempt to maximize their joint real income, and that the

---

[24] Mahoney, 1961, pp. 11.

[25] Samuelson, 1956, pp. 9 and pp. 21.

[26] For example, when the head of the family allocates his time to market work or leisure, the head must also consider how this allocation of time will have consequences on the utility of the rest of the family and on the household's joint income. See Becker, 1981, pp. 172-201. Further, it is a common observation that household members care for one another -- a decision to eat at a particular restaurant for example, takes into account of each household member's taste and preferences; planning vacation and concern for the safety of household members are other good examples.

household can thus be said to have a single family utility function.[27]

The second assumption that households only maximise is appealing since studies have shown that even in traditional societies, the responsiveness of the small tenant farmer to output price changes do not differ significantly from those of large commercial enterprises (the high income farmer).[28] There is thus no persuasive reason as to the contrary.

Finally, the criticism that the household production function requires constant returns to scale or at least late diminishing returns to scale appears reasonably valid. But at the same time, it must be recognised that a prior, we cannot rule out constant returns to scale totally. This is especially true since household production cannot be considered as operating within a fairly large scale typical of commercial firms. And as argued, the use of time as a measure of household production has its advantages. To state otherwise, empirical work would be severely limited.

In conclusion, the chapter has presented a method for the valuation of household production consistent in the way market goods and services are valued for social accounting purposes. It has been argued that household production consists of two components: MHP and NMHP which are identifiable and measurable. The next chapter shows how with the knowledge of the model and method of estimation described here can be used to estimate the value of household production in Singapore.

---

[27] Becker, 1981, pp. 172-185.

[28] See Evenson, 1978.

# 6 Household production in Singapore: An empirical study

The preceding chapters have laid the theoretical foundation behind household production measurement and valuation. In this chapter, we attempt to derive the quantity and economic value of household production in Singapore using the model and methods of estimation described earlier.

The chapter is organised as follows: The household survey is first briefly described and following this, a cross-sectional analysis involving descriptive statistics of household production patterns by different types and characteristics of households is made. The results are then used as the basis for deriving the value of household production in Singapore.

### 6.1 The household survey

The study was conducted over a 4-month period -- beginning February 1986 and ending May 1986 -- involving some one thousand households in Singapore, and more than thirty student interviewers[1]. The sampling frame is the master list of houses maintained and updated to September 1985 by the Department of Statistics. The Department of Statistics and the Ministry of Labour were consulted on sample selection.

In addition to obtaining demographic information on each of the family-household, household members were asked how many hours a week on average do they spend on each of the household tasks, with each of the household tasks -- cooking and meal preparation, after-meal clean-up, cleaning or regular house-care, laundry including ironing, accounts or matters of paper work, supervision of household tasks, shopping for

---

[1] Before this main survey was launched, a pilot survey was conducted to test the suitability of the questionnaire.

groceries, outdoor work or gardening, caring for one's children under 18 years of age, and the home education of one's children -- carefully defined. For those households with paid domestic help, they were also asked the total number of hours that were devoted to household work by their hired-help per se.

A contingent evaluation question[2] was then asked for those households with maids:

> If your household has to do all the chores by yourselves (instead of hiring domestic help to do part or all of the housework), then on average, how much time do you think your household would have to spend on doing all the chores per week? Please assume that the same amount and quality of housework had to be done per week.

Following the preceding section, knowledge of the total hours devoted to housework by each household inclusive and exclusive of paid domestic help, and response to the contingent evaluation question above enables us to calculate the value of the relative-efficiency coefficient, k.

Of the original one thousand households sampled, only 684 completed questionnaires were finally accepted and used for the survey analyses.

## 6.2      Descriptive statistics

The major results of the survey are grouped and discussed under the following sub-headings: patterns of time allocation according to size of household, effect of certain household characteristics on household production, household efficiency, and other results.

---

[2]  Contingent evaluation methods have been used, with growing success to value non-market goods and services. The technique, essentially involves the construction of a hypothetical market for the good in question and information is given as to its quantity, location, time dimensions and other attributes. The rules of operation of this contingent market are then established and the respondent is asked to indicate a reaction to some contingency that is posed. Contingent evaluation methods invariably use survey questionnaires to acquire the relevant data for analysis. While contingent evaluation methods to a large extent are found in the resource economic literature and used to price non-market goods, a first application was made to value household production for individual welfare purposes by Quah (1986: 875-889).

**Table 6.1**
**Average hours per week spent in household production**
**By all households and in families of different sizes**
**(Exclusive of hours done by maid)**

| Household Activities | Hours For Family of: | | | |
|---|---|---|---|---|
| | 2 Persons | 3-5 Persons | 6 or more Persons | Hours for all families |
| Cooking and meal preparation | 7.84 | 12.37 | 14.86 | 12.83 |
| After meal clean up | 2.68 | 4.11 | 4.35 | 4.09 |
| Cleaning or regular housecare | 4.63 | 6.13 | 6.17 | 6.04 |
| Laundry including ironing | 3.00 | 4.95 | 5.52 | 4.99 |
| Shopping for groceries | 3.57 | 4.99 | 5.59 | 5.08 |
| Outdoor work or gardening | 0.77 | 1.11 | 1.39 | 1.17 |
| Child care | 3.36[1] | 19.74[2] | 16.25[3] | 17.63[4] |
| Total (MHP) in hours | 25.85 | 53.40 | 54.13 | 44.46 |
| Accounts or matters of paperwork and supervision | 1.17 | 1.19 | 1.38 | 1.25 |
| Home education | 4.50[5] | 14.26[6] | 11.33[7] | 13.09[8] |
| Total (NMHP) in hours | 5.67 | 15.45 | 12.71 | 11.28 |
| HP (MHP + NMHP) in hours | 31.52 | 68.85 | 66.84 | 55.74 |
| n | 44 | 432 | 208 | 684 |

[1] n = 37    [2] n = 394    [3] n = 199    [4] n = 630
[5] n = 12    [6] n = 334    [7] n = 163    [8] n = 509

Tables 6.1 to 6.5 show the average hours per week spent in household production by all households, inclusive and exclusive of paid domestic help and in families of different sizes. In Table 6.1, the average time spent in household production per week by all households (exclusive of paid domestic help's time) and in families of different sizes are presented.[3] The results of the study showed that the average time spent at all household tasks, labelled HP in hours per week is 31.52 for a family of 2 persons, 68.85 for a family of 3-5 persons and 66.84 for a family of size 6 or greater. The average time spent for a family of any size is 55.74 hours per week.

Recall from Chapter 5, that HP activities comprise of MHP (market-replaceable household production) and NMHP (near-market replaceable household production) activities. From Table 6.1, the MHP hours per week for families of size 2, 3-5 and 6 or more are respectively, 25.85, 53.40 and 54.13; while the MHP hours per week for a family of any size is about 44.46. It might be noted that while the difference in MHP hours between a family of size 3-5 and a family of size 6 or more is only marginal, the difference is larger once the time spent in child care is removed. This difference is as large as 4.22 hours per week (see Table 6.1).

The average time spent in NMHP activities by all households (exclusive of paid domestic help's time) and in families of different sizes are respectively 5.67, 15.45 and 12.71 for families of size 2, 3-5 and 6 or more; while the average NMHP is 11.28 hours per week for a family of any size.

---

[3]   A weighted average is used in this and other tables where the time spent by domestic help is excluded. This is because of the lower averages of reported time spent in household production by households with maids (since we exclude the time spent by domestic help in these households). Essentially, the weighted average $(X_{1,2})$ took the form:

$$X_{1,2} = \frac{517}{684(X_1)} + \frac{167}{684(X_2)}$$

where there are 517 households without maids and, 167 households with maids in the sample of 684 households. $X_1$, $X_2$ are the averages of the time spent in household production by households without maids and with maids respectively.

The results of the survey also showed that, with the exception of families of size 2, child care appears to consume the most time vis a vis all other MHP household tasks. Removing the hours spent on child care reveals that cooking and meal preparation takes up the most time for families of all sizes. The weekly time spent in cooking and meal preparation averages 7.84, 12.37, and 14.86 hours for families of size 2, 3-5 and 6 or more respectively. This time spent is at least 50 percent more than the next highest time consumption household task, that is cleaning or regular housecare for families of size 3-5 and 6 or more.

Thus, a rank order of time commitment to the various MHP household tasks from highest to lowest indicates the following: child care, cooking and meal preparation, cleaning or regular housecare, shopping for groceries, laundry including ironing, after meal clean-up, and outdoor work or gardening. This pattern of time allocation to the various MHP household tasks is true for all family sizes whether inclusive or exclusive of paid domestic help (see Tables 6.1 to 6.5).

Table 6.4 shows that for those households with paid domestic help, their members continue to engage in household production, although the time allocated for most of the MHP household tasks is small with one exception; that being the time spent on child care which is 11.67, 16.09 and 16.27 for families of size 2, 3-5 and 6 or more, respectively.

Comparing Table 6.4 and Table 6.5, that is households with maids, exclusive and inclusive of paid domestic help's time, total MHP hours have risen from 28.72, 34.19 and 38.07 hours per week respectively for families of size 2, 3-5 and 6 or more to 56.36, 85.49 and 94.64 hours respectively. But unlike the rise in MHP hours, the NMHP hours have remained relatively stable excluding or including the paid domestic help's time. This confirms the a priori statement (see Chapter 1) that NMHP activities are not usually replaceable by hiring domestic help. It is often the case that hired domestic help simply do not engage in home management activities and the home education of their employer's children.

Further, in Tables 6.1 to 6.5, it can be seen that time used in MHP activities varied systematically according to the size of the household, with increasingly more time spent on household work as the size of the household increases. However, the same could not be said of NMHP activities. It appears that the NMHP hours for households of size 3-5 are larger than the NMHP hours for households of size 6 or more. In Table 6.1 for example, the NMHP hours for households with 3-5 persons averaged 15.45, while for households with 6 or more persons, the NMHP

148

hours are only about 12.71 hours, a difference of 2.74 hours a week. This may be explained by the presence of more children of school age in households of size 3-5 than in households of size 6 or more, and since one of the two constituents of NMHP activities is the home education of children, it can be expected that with more children of school age, more time is spent with them.[4]

However, on closer examination, there appears to be a difference in time spent in NMHP activities by households with and without paid domestic help. In Table 6.3, for example, which presents the average hours spent in household production per week by households without paid domestic help and in families of different sizes, the difference in NMHP hours between families of size 3-5 and 6 or more is almost 5.33 hours a week. This difference is due, in large part, to the difference in time spent in home education of children (12.54 hours for families of size 3-5 and 7.13 hours for families of size 6 or more).

However, for households with paid domestic help, the opposite results are observed. It appears that the NMHP hours are higher for families of size 6 or more than in families of size 3-5. This can be seen from Tables 6.4 and 6.5. The results also indicate that the time spent in both accounts and supervision, and in the home education of children is higher for families of size 6 or more than for those families of size 3-5. This is perhaps due to the fact that since much of the time spent in MHP activities is attributed to paid domestic help, family members can spend more time with their children and in the activities related to home organisation and management.[5]

---

[4] A check on the other component activity of NMHP activities, namely, accounts and supervision showed not much differences in time spent between families of size 3-5 and 6 or more, with the latter spending slightly more hours in this activity. See Tables 6.1 to 6.5.

[5] Even though they might have fewer children of schooling age.

149

## Table 6.2
### Average hours per week spent in household production
### By all households and in families of different sizes
### (Inclusive of hours done by maid)

| Household Activities | Hours For Family of: | | | |
|---|---|---|---|---|
| | 2 Persons | 3-5 Persons | 6 or more Persons | Hours for all families |
| Cooking and meal preparation | 9.25 | 14.79 | 17.64 | 15.30 |
| After meal clean up | 2.93 | 5.59 | 5.91 | 5.52 |
| Cleaning or regular housecare | 6.50 | 8.95 | 10.39 | 9.22 |
| Laundry including ironing | 4.25 | 6.28 | 7.88 | 6.63 |
| Shopping for groceries | 3.59 | 5.10 | 5.68 | 5.18 |
| Outdoor work or gardening | 0.90 | 1.64 | 2.08 | 1.73 |
| Child care | $1.16^1$ | $22.37^2$ | $19.25^3$ | $20.13^4$ |
| Total (MHP) in hours | 28.58 | 64.70 | 68.83 | 54.04 |
| Accounts or matters of paperwork and supervision | 1.17 | 1.19 | 1.38 | 1.25 |
| Home education | $4.50^5$ | $14.26^6$ | $11.33^7$ | $13.09^8$ |
| Total (NMHP) in hours | 5.50 | 15.62 | 12.72 | 11.28 |
| HP (MHP + NMHP) in hours | 34.08 | 80.32 | 81.55 | 65.32 |
| n | 44 | 432 | 208 | 684 |

[1] $n = 37$    [2] $n = 394$    [3] $n = 199$    [4] $n = 630$
[5] $n = 12$    [6] $n = 334$    [7] $n = 163$    [8] $n = 509$

150

## Table 6.3
### Average hours per week spent in household production by households without paid domestic help and in families of different sizes

| Household Activities | Hours For Family of: | | | |
|---|---|---|---|---|
| | 2 Persons | 3-5 Persons | 6 or more Persons | Hours for all families |
| Cooking and meal preparation | 8.74 | 13.83 | 17.13 | 14.46 |
| After meal clean up | 2.66 | 4.84 | 4.95 | 4.73 |
| Cleaning or regular housecare | 4.86 | 7.30 | 7.36 | 7.15 |
| Laundry including ironing | 3.49 | 6.13 | 7.15 | 6.25 |
| Shopping for groceries | 3.69 | 5.43 | 6.16 | 5.53 |
| Outdoor work or gardening | 0.46 | 0.90 | 1.05 | 0.92 |
| Child care | $1.23^1$ | $20.87^2$ | $16.24^3$ | $18.18^4$ |
| Total (MHP) in hours | 25.13 | 59.30 | 60.04 | 48.16 |
| Accounts or matters of paperwork and supervision | 1.00 | 1.18 | 1.26 | 1.19 |
| Home education | $3.62^5$ | $12.54^6$ | $7.13^7$ | $10.63^8$ |
| Total (NMHP) in hours | 4.62 | 13.72 | 8.39 | 8.91 |
| HP (MHP + NMHP) in hours | 29.75 | 73.02 | 68.43 | 57.07 |
| n | 35 | 350 | 152 | 517 |

[1] n = 35  [2] n = 330  [3] n = 152  [4] n = 517
[5] n = 11  [6] n = 263  [7] n = 122  [8] n = 396

## Table 6.4
### Average hours per week spent in household production
### By households with paid domestic help and in families of different sizes
### (Exclusive time spent by domestic help)

| Household Activities | Hours For Family of: | | | |
|---|---|---|---|---|
| | 2 Persons | 3-5 Persons | 6 or more Persons | Hours for all families |
| Cooking and meal preparation | 4.33 | 7.63 | 8.70 | 7.81 |
| After meal clean up | 2.78 | 1.72 | 2.74 | 2.12 |
| Cleaning or regular housecare | 3.72 | 2.34 | 2.95 | 2.62 |
| Laundry including ironing | 1.11 | 1.08 | 1.08 | 1.08 |
| Shopping for groceries | 3.11 | 3.56 | 4.03 | 3.69 |
| Outdoor work or gardening | 2.00 | 1.77 | 2.30 | 1.96 |
| Child care | $11.67^1$ | $16.09^2$ | $16.27^3$ | $15.91^4$ |
| Total (MHP) in hours | 28.72 | 34.19 | 38.07 | 35.19 |
| Accounts or matters of paperwork and supervision | 1.83 | 1.25 | 1.69 | 1.43 |
| Home education | $14.00^5$ | $20.63^6$ | $23.88^7$ | $21.76^8$ |
| Total (NMHP) in hours | 15.83 | 21.88 | 25.57 | 23.19 |
| HP (MHP + NMHP) in hours | 44.55 | 56.07 | 63.64 | 54.75 |
| n | 9 | 101 | 56 | 166 |

[1] n = 9   [2] n = 64   [3] n = 47   [4] n = 120
[5] n = 1   [6] n = 71   [7] n = 41   [8] n = 113

## Table 6.5
## Average hours per week spent in household production
## By households with paid domestic help and in families of different sizes
## (Inclusive time spent by domestic help)

| Household Activities | Hours For Family of: | | | |
|---|---|---|---|---|
| | 2 Persons | 3-5 Persons | 6 or more Persons | Hours for all families |
| Cooking and meal preparation | 11.22 | 17.90 | 19.05 | 17.93 |
| After meal clean up | 4.00 | 8.05 | 9.54 | 7.99 |
| Cleaning or regular housecare | 12.89 | 14.26 | 18.63 | 15.66 |
| Laundry including ironing | 7.22 | 6.72 | 9.84 | 7.79 |
| Shopping for groceries | 3.22 | 4.04 | 4.33 | 4.09 |
| Outdoor work or gardening | 3.14 | 4.44 | 5.29 | 4.67 |
| Child care | 14.67[1] | 30.08[2] | 28.96[3] | 29.08[4] |
| Total (MHP) in hours | 56.36 | 85.49 | 94.64 | 87.21 |
| Accounts or matters of paperwork and supervision | 1.83 | 1.25 | 1.69 | 1.43 |
| Home education | 14.00[5] | 20.62[6] | 23.87[7] | 21.75[8] |
| Total (NMHP) in hours | 15.83 | 21.87 | 25.56 | 21.09 |
| HP (MHP + NMHP) in hours | 72.19 | 107.36 | 120.20 | 95.03 |
| n | 9 | 101 | 56 | 166 |

[1] n = 9    [2] n = 64    [3] n = 47    [4] n = 120
[5] n = 1    [6] n = 71    [7] n = 41    [8] n = 113

In sum, with regard to patterns in time allocation in household production according to size of household, the following points are observed:

1    that for families of size 2, 3-5, and 6 or more persons, household production (exclusive of paid domestic help's time) is about 32 hours, 69 hours and 67 hours per week respectively.

2    that the MHP component of household production per week (exclusive of paid domestic help's time) for families of size 2, 3-5, and 6 or more persons is about 26 hours, 53 hours, and 54 hours respectively. That the MHP hours tend to vary systematically in an upward direction in that the larger the size of the household, the more hours are spent on MHP activities. This is true for both households with paid domestic help and for households without paid domestic help.

3    that the NMHP component of household production per week for families of size 2, 3-5, and 6 or more persons is about 6 hours, 15 hours, and 13 hours respectively. Unlike the rise in MHP hours when the time of a hired domestic help is included, the NMHP hours have remained relatively stable. This indicates that NMHP activities are not normally replaceable through the hiring of domestic help and this validates the a priori suggestion that MHP and NMHP can be treated as two different sets of activities in household production.

4    excluding NMHP activities, it is found that the most time-consuming activities in descending order are child care, cooking and meal preparation, regular housecare or cleaning, grocery shopping, laundry work and ironing, after meal clean-up, and gardening or outdoor-work. Including NMHP activities, it is found that the time spent in home education of children would rank about the same as the time spent in cooking and meal preparation but less than the time devoted to child care. The time spent in home management tends to be relatively stable (approximately less than 2 hours per week) across all household sizes, and would rank among the least time spent activity in all households.

*Effects of other household characteristics on household production*

## 1    Number of children and age group

The effect of the number of children on household production is shown in Tables 6.6 to 6.9. The results show that households with at least 1 child spend on average more than twice the time spent by households without children in household production. The time spent in household production also varies with the number of children a household has in that the more children, the higher the time spent in household production.

Thus, excluding the time spent by hired-help, the results indicate that on average, a household with 1 to 2 children spends 49.94 hours a week on MHP activities, 13.64 hours on NMHP, making a total of 63.58 hours of household production a week. In contrast, a household with 3-5 children spends 60.70 hours a week on MHP activities, 16.67 hours on NMHP activities, making a total of 77.37 hours of household production a week. Note that households without children spend on average a mere 30.29 hours a week on MHP activities, 1.30 on NMHP activities (only on home management activities since they do not have children), making a total of 31.59 hours of household production a week (See Table 6.6).

The results also show that households with pre-primary and primary school children spend even more hours in household production (10 to 20 hours more). This seems to imply that age of children may be an important factor in household production variation in that the younger the age of the child, the higher will be the time spent in household production. This result is not surprising since we would expect that more time and attention would be given to younger children than older children since the latter would presumably, to large part, be able to take care of themselves.

Tables 6.9, 6.10 and 6.11 are similarly interpreted. For example, looking at households with maids and including the time spent by hired-help, we find that on average, they spend more time in household production than those households without maids.

The greatest difference between families with children and those without was in the amount spent on child care but the time required for other household activities such as cooking, laundering, and cleaning also increased considerably the more children there are in the home.[6]

---

[6]    This assertion resulted from a check on the time spent on the individual tasks performed between households with children and those without (not shown here).

## Table 6.6
## Effect of the number and age of children on hours spent per week on household production
## ALL HOUSEHOLDS (excluding time spent by domestic help)

| | Households Without children (less than 18 years of age) 0 | Households with children (less than 18 years of age) 1-2 | 3-5[1] | Households with pre-primary school children 0 | 1 | 2 or more | Households with pre-primary school children (any number) |
|---|---|---|---|---|---|---|---|
| MHP | 30.29 | 49.94 | 60.70 | 46.57 | 65.72 | 66.57 | 63.34 |
| NMHP: | | | | | | | |
| Home management | 1.31 | 1.20 | 1.24 | 1.24 | 1.22 | 1.36 | 1.30 |
| Home education | - | 12.44 | 15.42 | - | 17.47 | 18.59 | 16.91 |
| Total | 1.31 | 13.64 | 16.66 | 1.24 | 18.69 | 19.95 | 18.21 |
| HP (MHP + NMHP) | 31.60 | 63.58 | 77.36 | 47.81 | 84.41 | 86.52 | 81.55 |
| n | 164 | 396 | 123 | 468 | 140 | 75 | 344 |

[1] Because there is only 1 household in the sample with 6 or more children, the reported hours spent in household production for this household have been omitted in this and other tables.

## Table 6.7
### Effect of the number and age of children on hours spent per week on household production:
### Households with maids
### (excluding time spent by domestic help)

| | Households without children (less than 18 years of age) | Households with children (less than 18 years of age) | | Households with pre-primary and primary school children (any number) |
|---|---|---|---|---|
| | 0 | 1-2 | 3-5 | |
| MHP | 18.22 | 32.96 | 44.67 | 36.89 |
| NMHP | 1.55 | 23.60 | 22.81 | 25.57 |
| HP(MHP+NMHP) | 19.77 | 56.56 | 67.37 | 62.46 |
| n | 42 | 90 | 34 | 85 |

## Table 6.8
### Effect of the number and age of children on hours spent per week on household production:
### Households without maids

| | Households without children (less than 18 years of age) | Households with children (less than 18 years of age) | | Households with pre-primary and primary school children (any number) |
|---|---|---|---|---|
| | 0 | 1-2 | 3-5 | |
| MHP | 34.46 | 55.17 | 66.83 | 72.03 |
| NMHP | 1.22 | 11.02 | 14.48 | 15.94 |
| HP(MHP+NMHP) | 35.68 | 66.19 | 81.31 | 87.97 |
| n | 122 | 306 | 89 | 259 |

**Table 6.9**
**Effect of the number and age of children on hours spent**
**per week on household production:**
**Households with maids**
**(including time spent by domestic help)**

| | Households without children (less than 18 years of age) | Households with children (less than 18 years of age) | | Households with pre-primary and primary school children (any number) |
|---|---|---|---|---|
| | 0 | 1-2 | 3-5 | |
| MHP | 53.10 | 84.77 | 96.58 | 100.58 |
| NMHP | 1.56 | 24.11 | 23.05 | 26.15 |
| HP(MHP+NMHP) | 54.66 | 108.88 | 119.63 | 126.73 |
| n | 42 | 90 | 34 | 85 |

In sum, the results show that families with children did more household production than those without, and that the larger the number, and the younger the age of children, the higher the time spent on household production.

## 2    Wives' Employment Status

Tables 6.10 and 6.11 show respectively the effects of wives' employment status on time spent on household production in hours per week excluding and including the time spent by paid domestic help. Excluding the time spent by domestic help, the results show that households with employed wives spend on average 44.02 hours on MHP activities, 17.32 hours on NMHP activities, making a total of 61.34 hours a week on household production. However, households with unemployed wives spend on average 56.94 hours on MHP activities, 12.34 hours on NMHP activities, making a total of 69.28 hours a week.

Thus, the results show that households with unemployed wives spend on average more hours in household production than households with employed wives. The same result is also observed for households with maids including the time spent by hired help (Table 6.14), while they are

158

about the same for households with maids excluding the time spent by hired help (Table 6.13).

While a large part of the time spent in household production can be attributed to MHP activities for both households with employed and unemployed wives, and that the former is less than the latter, just the opposite results are observed for the time spent in NMHP activities in that unemployed-wife households spend on average less time in NMHP activities than employed-wife households. This is true for all cases including and excluding the time of paid domestic help, and for households with and without domestic help.

It may be the case that employed-wife households are relatively higher educated than unemployed-wife households and as such we would expect the former to spend more time in the home education of their children.[7] The fact that unemployed-wife households spend more time on average on MHP activities than it is the case with employed-wife households is simply the result of more home time being available to unemployed-wife households.

## 3   Level of schooling of husband and wife

Tables 6.15 and 6.16 present the results of the effect of the level of schooling of husband and wife on a family's household production in hours per week excluding and including the time spent by paid domestic help respectively.

Excluding the time spent by domestic help, it appears that where the wife or husband's education is at least a post-secondary qualification, the time spent in household production for these families is far less than that of families whose wife or husband's education is a primary or secondary school qualification. This disparity is even higher when considering MHP activities alone. In fact, it is the time spent on MHP activities that accounts for much of the differences between the former and the latter groups.

In the case of families where wife and husband's education are that of no schooling, the results indicate that their time spent in MHP activities is closer to those families with primary school or secondary school educated wife and husband. The disparity in total household production time between households with uneducated wives or husbands and that of

---

[7]   A check on the hours spent on home management showed no difference between employed and unemployed wife households.

159

households with wives or husbands holding primary or secondary qualifications lies not so much on time spent on MHP activities but more so on time spent on NMHP activities. In Table 6.15 for example, the results show that households of the former spend on average only 7.56 to 10.18 hours a week on NMHP activities whereas households of the latter spend between 13 to 19 hours on NMHP activities. These results are also observed in Table 6.16 where the time of paid domestic help is included.

**Table 6.10**
**Effect of wives' employment status on time**
**spent in household production in hours per week**
**ALL HOUSEHOLDS (excluding time spent by domestic help)**

|  | MHP | NMHP | HP(MHP+NMHP) |
|---|---|---|---|
| Households with employed wives n1 | 44.02 272 | 17.32 | 61.34 |
| Households with unemployed wives n2 | 56.94 412 | 12.34 | 69.28 |

**Table 6.11**
**Effect of wives' employment status on time**
**spent in household production in hours per week**
**ALL HOUSEHOLDS (including time spent by domestic help)**

|  | MHP | NMHP | HP(MHP+NMHP) |
|---|---|---|---|
| Households with employed wives n1 | 58.57 272 | 17.47 | 76.04 |
| Households with unemployed wives n2 | 67.03 412 | 12.41 | 79.44 |

160

**Table 6.12**
**Effect of wives' employment status on time spent**
**in household production in hours per week: households with maids**

|  | MHP | NMHP | HP(MHP+NMHP) |
|---|---|---|---|
| Households with employed wives n1 | 47.65 187 | 14.12 | 61.77 |
| Households with unemployed wives n2 | 65.59 330 | 10.58 | 76.17 |

**Table 6.13**
**Effect of wives' employment status on time spent**
**in household production in hours per week: households with maids**
**(excluding time spent by domestic help)**

|  | MHP | NMHP | HP(MHP+NMHP) |
|---|---|---|---|
| Households with employed wives n1 | 36.03 85 | 23.84 | 59.87 |
| Households with unemployed wives n2 | 34.26 82 | 22.23 | 56.49 |

## Table 6.14
### Effect of wives' employment status on time spent in household production in hours per week: households with maids (including time spent by domestic help)

|  | MHP | NMHP | HP(MHP+NMHP) |
|---|---|---|---|
| Households with employed wives n1 | 86.06 85 | 24.33 | 110.39 |
| Households with unemployed wives n2 | 88.42 82 | 22.59 | 111.01 |

## Table 6.15
### Household production and level of schooling of husband and wife in hours per week
### ALL HOUSEHOLDS (excluding time spent by domestic help)

|  | Wife's Education | | | | |
|---|---|---|---|---|---|
|  | No schooling | Primary school | Secondary school | Post secondary | Polytechnic and University |
| MHP | 52.44 | 56.56 | 57.79 | 44.33 | 38.08 |
| NMHP | 7.56 | 13.22 | 19.23 | 18.06 | 22.36 |
| HP(MHP+NMHP) | 60.00 | 69.78 | 77.02 | 62.39 | 60.44 |
| n | 194 | 183 | 161 | 56 | 60 |
|  | Husband's Education | | | | |
| MHP | 51.72 | 56.77 | 62.75 | 44.39 | 40.44 |
| NMHP | 10.18 | 18.85 | 19.22 | 14.19 | 23.64 |
| HP(MHP+NMHP) | 61.90 | 75.62 | 81.97 | 58.58 | 64.08 |
| n | 87 | 192 | 171 | 43 | 114 |

**Table 6.16**
**Household production and level of schooling of**
**husband and wife in hours per week**
**ALL HOUSEHOLDS (including time spent by domestic help)**

|  | Wife's Education | | | | |
|---|---|---|---|---|---|
|  | No school-ing | Primary school | Secondary school | Post secondary | Polytechnic and University |
| MHP | 53.59 | 60.57 | 73.06 | 73.72 | 91.44 |
| NMHP | 7.57 | 13.28 | 19.38 | 18.18 | 23.07 |
| HP(MHP+NMHP) | 61.16 | 73.85 | 92.44 | 91.90 | 114.51 |
| n | 194 | 183 | 161 | 56 | 60 |
|  | Husband's Education | | | | |
| MHP | 51.82 | 59.69 | 71.09 | 65.42 | 85.89 |
| NMHP | 10.21 | 18.87 | 19.27 | 14.22 | 24.01 |
| HP(MHP+NMHP) | 62.03 | 78.56 | 90.36 | 79.64 | 110.06 |
| n | 87 | 192 | 171 | 43 | 114 |

**Table 6.17**
**Household production and education grouping in hours per week**
**ALL HOUSEHOLDS (Excluding time spent by domestic help)**

|  | Both Husband and wife in High Education Group | Both Husband and Wife in Low Education Group | One spouse in High Education and the Other in Low Education Group |
|---|---|---|---|
| MHP | 38.15 | 61.55 | 47.45 |
| NMHP: | | | |
| Home Management | 1.58 | 1.24 | 1.52 |
| Home Education | 21.83 | 12.05 | 21.60 |
| Total | | | |
| HP(MHP+NMHP) | 61.56 | 74.84 | 70.57 |
| n | 46 | 376 | 42 |

**Table 6.18**
**Effect of household income on household production in hours per week**
**ALL HOUSEHOLDS (Excluding time spent by domestic help)**

| Income Groups ($) Per Month | n | MHP | NMHP | HP(MHP + NMHP) |
|---|---|---|---|---|
| Below 500 | 38 | 59.88 | 6.94 | 66.82 |
| 500-999 | 156 | 56.92 | 11.59 | 68.51 |
| 1000-1499 | 125 | 53.05 | 10.66 | 63.71 |
| 1500-1999 | 74 | 60.01 | 13.28 | 73.29 |
| 2000-2499 | 67 | 66.10 | 19.87 | 85.97 |
| 2500-2999 | 39 | 48.71 | 17.97 | 66.6 |
| 3000 and above | 179 | 36.81 | 19.59 | 56.40 |

**Table 6.19**
**Effect of household income on household production in hours per week**
**ALL HOUSEHOLDS (Including time spent by domestic help)**

| Income Groups ($) Per Month | n | MHP | NMHP | HP(MHP + NMHP) |
|---|---|---|---|---|
| Below 500 | 38 | 59.89 | 6.94 | 66.83 |
| 500-999 | 156 | 57.02 | 11.61 | 68.63 |
| 1000-1499 | 125 | 54.59 | 10.67 | 65.26 |
| 1500-1999 | 74 | 61.81 | 13.29 | 75.10 |
| 2000-2499 | 67 | 70.55 | 20.01 | 90.56 |
| 2500-2999 | 39 | 66.56 | 17.68 | 84.24 |
| 3000 and above | 179 | 74.62 | 19.97 | 94.59 |

165

In fact, the results reveal that families with uneducated wives or husbands spent the least amount of time in NMHP activities compared to families whose wives or husbands belong to other educational categories. This disparity is even more marked when households with wives or husbands holding at least post-secondary qualifications are compared. The difference is as large as twice in the case of comparing husband's education and three times in the case of comparing wife's education (Tables 6.15 and 6.16).

To provide a better summary of the effect of spouses' education on household production, Table 6.17 classified educational attainment of the spouses into three educational groups: both husband and wife in high education group, defined as holders of post-secondary, polytechnic or university qualifications; both husband and wife in low education group, defined as holders of secondary and primary school qualifications, and the uneducated; and one spouse in high education and the other in low education group. The difference between Tables 6.15, 6.16 and 6.17 is that the latter is more restrictive in that both spouses' education are taken into consideration whereas in the former, only the education level of one of the spouses and the effect on household production is considered.

The results show that household production is highest (74.84 hours a week) in families where both husband and wife are in the low education group, excluding the time spent by domestic help. Households where bothspouses are in the high education group did the least amount of household production on average (61 hours a week or about 10 hours less than the other two education groups). There appears however to be not much difference in time spent in household production between families where both husband and wife are in the low education group, and husband and wife of mixed educational group (that is, one spouse in high education group and the other in low). On closer examination, the results indicate that while households of the low education group spend on average 14 hours more a week in MHP activities over the mixed education group, this difference is largely offsetted by an equally large difference between the two groups in time spent in NMHP activities in that households of the mixed education group spend about 9 hours more in home education of children than households of the low education group.

Tables 6.18 and 6.19 show respectively the effects of household income on time spent on household production in hours per week excluding and including time spent by paid domestic help.

Excluding time spent by domestic help, the results show that households with an income bracket of $2000 to $2499 did the most household production (85.97 hours), and this is also the case for time spent in MHP activities (66.10 hours) and NMHP activities (19.87). The next highest group in terms of time devoted to household production is the income bracket $1500-$1999 performing 73.29 hours a week. The rest of the households with other income brackets (except $3000 and above) did more or less the same amount of housework. Households with an income bracket $3000 and above did the least amount of household production (56.40 hours) and the least amount in MHP activities; but the second highest (and very close too) in terms of time devoted to NMHP activities (19.59 hours).

It is not surprising that households with the income bracket $3000 and above spend on average the least time in household production since presumably the majority of these households have maids and/or may eat out more often. When the time of paid domestic help is included (Table 6.19), the results show that it is households with the income bracket $3000 and above that spend the most time in household production. Apparently, much of the household production is carried out by paid domestic help for this income group. In Table 6.20, the effect of household income on household production for households without paid domestic help showed again that households with an income bracket $3000 and above did the least household production (57.53 hours a week) and that households with an income bracket $2000-$2499 did the most household production (85.87 hours a week).

Alternatively, the effect of household income on household production can be seen in terms of spousal occupation (and income derived thereof) and household production. Classifying spouses who are currently employed in education, managerial and professional services as the high income group, and spouses who are in production or manual and clerical services as the low income group, Table 6.21 shows the effect of income groups on time devoted to household production.

**Table 6.20**
**Effect of household income on household production in hours per week**
**Households Without Maids**

| Income Groups ($) Per Month | n | MHP | NMHP | HP(MHP+NMHP) |
|---|---|---|---|---|
| Below 500 | 38 | 59.89 | 6.94 | 66.83 |
| 500-999 | 154 | 57.11 | 11.43 | 68.54 |
| 1000-1499 | 120 | 54.27 | 10.56 | 64.83 |
| 1500-1999 | 70 | 61.43 | 13.35 | 74.78 |
| 2000-2499 | 56 | 68.23 | 17.64 | 85.87 |
| 2500-2999 | 21 | 58.26 | 11.81 | 70.07 |
| 3000 and above | 56 | 45.84 | 11.69 | 57.53 |

**Table 6.21**
**Household production and income grouping in hours per week**
**ALL HOUSEHOLDS (Excluding time spent by domestic help)**

| | Both husband and wife in high income group | Both husband and wife in low income group | One member in high income and the other in low income group |
|---|---|---|---|
| MHP | 38.65 | 50.39 | 47.19 |
| NMHP | 19.14 | 14.31 | 13.78 |
| HP(MHP+NMHP) | 57.79 | 64.70 | 60.97 |
| n | 66 | 99 | 51 |

168

**Table 6.22**
**Home production and income grouping in hours per week**
**Households Without Maids**

|  | Both husband and wife in high income group | Both husband and wife in low income group | One member in high income and the other in low income group |
|---|---|---|---|
| MHP | 39.25 | 50.54 | 58.54 |
| NMHP | 14.43 | 17.21 | 15.19 |
| HP(MHP+NMHP) | 53.68 | 67.75 | 73.73 |
| n | 11 | 95 | 29 |

**Table 6.23**
**Household production and income grouping in hours per week**
**ALL HOUSEHOLDS (Inclusive of time spent by domestic help)**

|  | Both husband and wife in high income group | Both husband and wife in low income group | One member in high income and the other in low income group |
|---|---|---|---|
| MHP | 78.28 | 51.13 | 72.01 |
| NMHP | 24.36 | 18.16 | 16.74 |
| HP(MHP+NMHP) | 103.64 | 69.29 | 88.75 |
| n | 66 | 99 | 51 |

Excluding the time spent by domestic help, the results show that households where both husband and wife are in the high income group, spend on average, the least amount of time in household production (57.79 hours a week) whereas households where spouses are both in the low income group spend the most amount of time (64.70 hours). The mixed income group where one spouse member belongs to a high income group and the other in the low income group shows a household production effort of about 60.97 hours a week. Again, time spent on NMHP activities is highest for the high income group.

The same results are also observed in the case of households which do not hire domestic help in that households of the high income group spend on the average the least time in household production (Table 6.22). However, when the time of paid domestic help is included, the results are again reversed in that households of the high income group did the most household production (Table 6.23).

*Household efficiency*

The relative efficiency coefficient measures the relative efficiency in performing market replaceable household production by family-household members vis a vis paid domestic help. Recall from Chapter 5 that the relative efficiency is called k and essentially it gives a rough idea of the difference in time spent in MHP activities in hours per week between family members engaged in MHP activities and non-family members in the form of hired domestic help performing the same amount or physical load of household production. Thus, a value of k > 1 implies that the paid domestic help is more efficient whereas a value of k < 1 means the household members are more efficient. When k approaches 1 or is equal to 1, the family member and the hired-help are said to be equally efficient.

The value of the relative efficiency coefficient, k is derived from the survey using questions 28, 29 and 18 (see Appendix). Without repeating much of Chapter 5, it suffices to say that the relative efficiency coefficient results from the households' responses (for those households with maids) of how much time it would take them to do the same amount of market replaceable household production as is now done in their household (Q28) inclusive of paid domestic help's effort. Since not all of the market replaceable household production would be done by hired-help as family members continue to do some of it (sum total of Q18), it is essential that in deriving the value of the relative efficiency coefficient, only the remaining portion which is now done by hired-help be considered (sum

170

total of Q29 - sum total of Q18).  So that,

$$k = \frac{Q28 - \sum Q18}{\sum Q29 - \sum Q18}$$

where $\Sigma$ = sum total of MHP household tasks in hours.

Table 6.24 presents some estimates of the relative efficiency coefficient by family size, number of children, education and occupation of wives and income group.  We note that a family of size 2 is the least efficient since the k value of 1.64 is the highest whilst a family of size 6 or more is the most efficient with the lowest k value of 1.33.  That a family of size 2 is the least efficient can be explained by the fact that most if not all of these households hire a 'specialised' domestic help to do only one or two household task(s) such as laundry and ironing.[8]  Since we would expect that the household tasks that are hired out would be those tasks which the family finds them either to have the most disutility (in terms of the act of performing them) and/or in which family members are the least efficient in producing them, the lower the household's efficiency it would be for the one or two household tasks that are hired out.  However, as more and more household tasks are delegated to a paid domestic help (as this is usually the case of larger sized families), we would expect the hired-help's efficiency to fall.[9]

---

[8]   The results of the survey showed that close to 85 percent of the households with maids and of family size 2 presently hire a domestic help to do only one or two activities, namely laundry and ironing, and regular cleaning.

[9]   The standard argument being one based on specialisation versus non-specialisation.

**Table 6.24**
**Relative efficiency coefficient by size of family, and number of children**
**(Less than 18 years of age)**
**Households with maid (n = 167)**

|  | Relative efficient coefficient |
|---|---|
| Size of family | |
| 2 | 1.64 |
| 3-5 | 1.46 |
| 6 or more | 1.33 |
| Number of children | |
| 0 | 1.62 |
| 1-2 | 1.43 |
| 3-5 | 1.40 |

**Table 6.25**
**Relative efficiency coefficient by education and occupation of wives**
**Households with maids (n = 167)**

|  | Relative efficient coefficient |
|---|---|
| Education of wives | |
| No education of schooling | 1.27 |
| Completed primary school | 1.38 |
| Completed secondary school | 1.36 |
| Completed post secondary school | 1.38 |
| Completed polytechnic/university | 1.37 |
| Occupation of wives | |
| Education | 1.70 |
| Managerial | 1.31 |
| Professional/Technical | 1.79 |
| Production/Transport/Manual | 1.34 |
| Sales/Services/Clerical | 1.13 |

Similar results are observed for estimates of the relative efficiency coefficient by number of children. Where number of children is small, the relative efficiency coefficient is high and where the number of children is large, the relative efficiency coefficient falls, although the survey results showed by not much (k = 1.43 for 1-2 children, and k = 1.40 for 3-5 children).[10] But for households without children, the relative efficiency coefficient appears to be relatively high; that is k = 1.62.

Table 6.25 presents some estimates of the relative efficiency coefficient by education and occupation of women spouses. The results show that wives with no education appear to be the most efficient in market replaceable household production (k = 1.27). Wives with primary, secondary, post secondary and tertiary education appear to be less efficient with k values of 1.38, 1.36, 1.38, and 1.37 respectively.

When the occupation of wives are considered the results show that wives who are in Education and Professional or Technical fields appear to be least efficient relative to those in managerial, production and sales, services or clerical. For those women spouses who are in sales, services or clerical, their household efficiency is highest with a k value of 1.13.

Finally, in Table 6.26, estimates of the relative efficiency coefficient by income groups are presented. It appears that households with an income bracket $1000 to $1499 a month are the most efficient having the lowest value of k = 1.14, whilst households with an income bracket $2500 to $2999 are the least efficient having the highest value of k = 1.84.

In sum, it might be observed that the relative efficiency coefficient by all the different household characteristics presented here from Tables 6.24 to 6.26 shows a positive and greater than one value of k meaning that in general, paid domestic help is more efficient in market replaceable household production than family members. These results are not surprising particularly since the majority of the paid domestic help come from the relatively less developed foreign countries in this region such as Philippines, Sri Lanka and Indonesia where self-performed household production is a way of life and as such they may be said to have more practices than say Singaporean households. But even if this reason is not valid, we would also expect that paid domestic help would be more efficient on average than household members for the simple reason that

---

10    We omit households with 6 or more children since there is only one household with such a characteristic.

173

it is their job.[11]

**Table 6.26**
**Relative efficiency coefficient by income group**
**Household with maids (n = 167)**

| Income group ($) per month | Relative efficiency coefficient |
|---|---|
| Below 500 | -[1] |
| 500-999 | -[1] |
| 1000-1499 | 1.14 |
| 1500-1999 | 1.54 |
| 2000-2499 | 1.30 |
| 2500-2999 | 1.84 |
| 3000 and above | 1.36 |

[1]     Because there are 2 households only for these income groups, the reported relative efficiency values are necessarily omitted.

*Value of household production in Singapore*

Having determined in the earlier part of this chapter the average hours devoted to family-household production per week, the next step is to impute a monetary value on this time. To do this, we refer to Chapter 5 where the method of imputation for the value of household production is given by the efficiency-adjusted replacement cost method for MHP activities, and the replacement cost of itemized services method for NMHP activities.

---

[11]     For a discussion on other results and also a hypothesis testing on possible links between differences in the reported number of hours devoted to household tasks, please refer to Quah, 1987.

## 6.3    Source and methodology used

Essentially, for market replaceable household production (MHP) activities, this involves the multiplication of the reported hours devoted to household production by the average hourly wage rates for a full-time paid domestic help and reducing this dollar amount by the relative-efficiency coefficient, k (where in the case of Singapore, it is found that on average and by household characteristic, the paid domestic help is more efficient than the household member (s) in performing MHP activities).[12]   In accounting for the value of NMHP activities, where NMHP activities consist of home management and home education of children activities, we use the average hourly wage rate of managers of small firms, and the average hourly wage rate of primary and pre-primary school teachers respectively, each multiplied by their respective hours.[13]

The average monthly wage rate of paid domestic help is estimated by taking the arithmetic mean of the reported wage-rates of Filipino, Indonesian and Sri Lankan household maids from a survey of some 20 firms supplying these maids to Singapore households.   It has been estimated that there are more than 30,000 foreign maids in Singapore, mainly Filipino, Sri Lankan and Indonesia.[14]  Their popularity has grown since living standards in Singapore have risen significantly over the last decade and as such the largely affluent Singapore households could now afford to hire domestic help.   Further, foreign maids are desired by Singaporean households mainly because of their lower wage rate on average vis a vis the Singaporean maids, and the fact that most of these foreign maids tend to come from countries where they have helped out in the menial housework chores from an early age and as such are used to such kinds of work.

An estimated one in every twenty households in Singapore now has a maid and the recruitment drive for foreign maids has become more

---

[12]   See Chapter 6 section on *Household efficiency.*

[13]   See Chapter 5.

[14]   There are currently no published reports on the exact number of foreign maids in Singapore but an indication of this number is given by the Ministry of Labour, Work Permit Division.

intense among local firms supplying these maids.[15]

Unlike Western countries where hired domestic help is usually paid on a per hour basis, in Singapore they are usually paid on a monthly basis. Thus, the average hourly wage rate of paid domestic help has to be derived. This is done by dividing the average monthly wage rate of paid domestic help by their average monthly hours devoted to housework chores. The average monthly wage rate of paid domestic help is, as mentioned, derived from the reported wage rates of the three dominant groups of foreign maids supplied to Singapore households and this is calculated to be about $205.37. The household survey reveals that a full-time paid domestic help spends on average 7.48 hours a day or 44.87 hours a week on MHP activities. Thus, the average per hour wage rate of a paid full-time domestic help is estimated to about $1.15.

This average per hour wage rate of hired maids is then adjusted for efficiency differences in performing MHP activities between the homemaker and the hired-help. The relative-efficiency coefficient will be different depending on which household characteristic is chosen (for example, by size, number of children, or occupation of wives; see Tables 6.24 and 6.25).

To determine the hourly wage rates for NMHP activities, we must first recognise that there are two services included in the definition of NMHP activities: home management and home education. For services considered as home management -- planning menus, settlement of bills and other accounts, and the supervision of hired-help (if any) -- we use the average wage rate of a manager of a very small firm, and this is obtained from the Report on the Labour Force Survey of Singapore 1985 (Table 39, page 56).

Dividing this average monthly wage rate received by a manager of a very small firm by 176 hours (assuming that he or she works on a 5 day-week at 8 hours a day), the average per hour wage is estimated to be about $1.25. The reason for using the average wage rate of a manager of a very small firm and not say, the wage rate of a manager of a relatively larger firm is that in most cases, the average household size is not more than 6 persons with the maximum number being not more than 10 persons. The

---

[15]  Very little is written on foreign maids in Singapore and much of what is written here comes from the author's conversations with personnel of the Department of Statistics, and the Ministry of Labour. But see also, a report by Geoffrey Murray titled, "Singapore's Domestic Squabble" carried in a popular magazine called PHP Intersect, December 1986 pp. 36-67.

household can thus be compared to a very small firm -- less than say 15 employees -- producing goods and services not for market consumption however but for own consumption.

The other NMHP activity, namely the home education of children is valued by using the average wage rate of pre-primary and primary school teachers (obtained from the Ministry of Education) divided by 176 hours (again, assuming that he or she works on a 5 day week at 8 hours a day).[16] This then gives us the average wage rate per hour of a primary cum pre-primary school teacher at $3.52.

The calculation of the total value of household production then proceeds with a multiplication operation involving the homemaking hours estimated, relative-efficiency
coefficients, and the average hourly wage rates of a paid domestic help, manager of a very small firm, and primary cum pre-primary school teacher, and then summing them up.

### 6.4 Weekly and annual value of household production by household characteristics

We first present estimates of the weekly and annual value of household production for the sample of households used. Based on the reported weekly hours devoted to household production by families of different sizes (Table 6.1), Table 6.27 below derives the corresponding weekly and annual dollar estimates on household production for the sample of households used. While Table 6.27 excludes the dollar contributions made by paid domestic help, Table 6.28 includes them.

As can be seen from Table 6.27, the average weekly and annual dollar values of household production respectively amount to $35 and $1834 for a family of size 2, $93 and $4856 for a family of size 3-5, and $87 and $4576 for a family of size 6 or more persons. Clearly, the value of household production is highest for the family size of 3-5 persons. While the value of household production is higher for a family of 3-5 persons than that of a family of 6 or more persons, one will notice that much of

---

16 A school teacher in Singapore is normally required to attend extra-curricular activities on Saturdays and/or one other afternoon during a week-day. It is also estimated that he or she spends, on average, at least 2 hours each day on weekdays after-school hours attending to the marking of student essays and other assignments.

the incremental value of the former can be attributed to the higher value spent on home education; presumably the presence of more younger primary-going school children than that of the latter.

While Table 6.27 shows the average weekly and annual dollar values of household production excluding the contributions by paid domestic help, Table 6.28 includes the dollar value contributions of paid domestic help, and by different family sizes. As in Table 6.27, the results in Table 7.2 are similarly interpreted. An interesting note, however, is that, the contributions made by paid domestic help increase the value of MHP by almost $10 in the case of a family of 3-5 persons and about $13 in the case of a family of 6 or more persons. The contribution of paid domestic help appeared not to have changed much for a family of 2 persons, mainly because very rarely do households of this size hire domestic help.[17]

The weekly and annual dollar values of household production by household characteristics other than by size of the household are presented in Tables 6.29 (by number of pre-primary school children), 6.30 (by educational attainment of the wife), and 6.31 (by level of education of both spouses).

In households without pre-primary school children, the weekly and annual dollar values of household production amount to about $54 and $2841 respectively. The presence of one pre-primary child adds about $83 and $4331 to the weekly and annual value of household production respectively. The incremental value for 2 or more pre-primary children is about $5 and $264 per week and per year respectively. Thus, the value of household production is highest in families with 2 or more pre-primary school children. This is all shown in Table 6.29.

Table 6.30 shows the value of household production according to the level of education attained by the wife. It appears that while the value of household production is highest in families where the wife has completed a secondary education ($113 a week, and $5887 a year), the difference in value is insignificant with wives of other educational backgrounds except no schooling. For families where the wife had no schooling, the value of household production is lowest at $71 and $3698 per week and per year respectively. On closer examination, the lower value of household production for families with uneducated wives can be attributed to the lower dollar values assigned to the home education of children (about $20 and $1000 less per week and per year respectively than the next higher

---

[17] A check on the sample showed only 9 out of 684 households hire a domestic help for this household size.

178

group ie families with primary school educated wives, and more than $51 and $2650 per week and per year respectively less than families with university or polytechnic educated wives).

**Table 6.27**
**Average weekly and annual dollar value of household production**
**by size of family**
**(Excluding the contributions of paid domestic help)**

| Type of Household Activities | Per household annual value ($) Family size | | | |
|---|---|---|---|---|
| | 2 | 3-5 | 6 or more | Any size |
| MHP | 17.97 | 41.70 | 46.40 | 5.36 |
| NMHP: | | | | |
| Home management | 1.46 | 1.49 | 1.73 | 1.56 |
| Home education | 15.84 | 50.20 | 39.88 | 35.31 |
| HP(MHP+NMHP) | 35.27 | 93.39 | 87.21 | 72.23 |
| Type of Household Activities | Per household weekly value ($) Family size | | | |
| | 2 | 3-5 | 6 or more | Any size |
| MHP | 934.44 | 2168.40 | 2412.80 | 1838.72 |
| NMHP: | | | | |
| Home management | 75.92 | 77.48 | 89.96 | 81.12 |
| Home education | 823.68 | 2610.40 | 2073.76 | 1836.12 |
| HP(MHP+NMHP) | 1834.04 | 4856.28 | 4576.52 | 3755.96 |

**Table 6.28**
**Average weekly and annual dollar value of household production**
**by size of family**
**(Including the contributions of paid domestic help)**

| Type of Household Activities | Per household annual value ($) Family size | | | |
|---|---|---|---|---|
| | 2 | 3-5 | 6 or more | Any size |
| MHP | 17.87 | 50.52 | 59.00 | 43.13 |
| NMHP: | | | | |
| Home management | 1.25 | 1.70 | 1.74 | 1.69 |
| Home education | 15.84 | 50.20 | 39.88 | 35.31 |
| HP(MHP+NMHP) | 36.96 | 102.49 | 100.62 | 80.11 |

| Type of Household Activities | Per household weekly value ($) Family size | | | |
|---|---|---|---|---|
| | 2 | 3-5 | 6 or more | Any size |
| MHP | 1033.24 | 2627.04 | 3068.00 | 2242.76 |
| NMHP: | | | | |
| Home management | 65.00 | 88.40 | 90.48 | 87.88 |
| Home education | 823.68 | 2610.40 | 2073.76 | 1836.12 |
| HP(MHP+NMHP) | 1921.92 | 5325.84 | 5232.24 | 4160 |

## Table 6.29
## Average weekly and annual dollar value of household production by number of pre-primary school children
### (Excluding the contributions of paid domestic help)

| Type of Household Activities | Per household weekly value ($) Number of Pre-Primary School Children (Less than or equal to 6 years of age) | | |
|---|---|---|---|
| | 0 | 1 | 2 or more |
| MHP | 53.09 | 74.92 | 75.89 |
| NMHP: Home management | 1.55 | 1.53 | 1.70 |
| Home education | - | 61.49 | 65.44 |
| Total | 1.55 | 63.02 | 67.14 |
| HP(MHP+NMHP) | 54.64 | 137.94 | 143.03 |
| Type of Household Activities | Per household annual value ($) Family size | | |
| | 0 | 1 | 2 or more |
| MHP | 2760.68 | 3895.84 | 3946.28 |
| NMHP: Home management | 80.60 | 79.56 | 88.40 |
| Home education | - | 3197.48 | 3402.88 |
| Total | 80.60 | 3277.04 | 3491.28 |
| HP(MHP+NMHP) | 2841.28 | 7172.88 | 7437.56 |

## Table 6.30
## Average weekly and annual dollar value of household production
## by educational attainment of the wife
## (Excluding the contributions of paid domestic help)

| Type of Household Activities | Per household weekly value ($) | | | | |
|---|---|---|---|---|---|
| | No school-ing | Primary school | Secondary | Post Secondary | University and Polytechnic |
| MHP | 47.07 | 46.72 | 48.44 | 36.62 | 31.69 |
| NMHP: | | | | | |
| Home management | 1.43 | 1.46 | 1.63 | 1.65 | 1.81 |
| Home education | 22.63 | 42.45 | 63.15 | 58.92 | 73.64 |
| Total | 24.06 | 43.91 | 64.78 | 60.57 | 75.45 |
| HP(MHP+NMHP) | 71.13 | 90.63 | 113.22 | 97.19 | 107.14 |
| Type of Household Activities | Per household annual value ($) | | | | |
| | No school-ing | Primary school | Secondary | Post secondary | University and Polytechnic |
| MHP | 2447.64 | 2429.44 | 2518.88 | 1904.24 | 1647.88 |
| NMHP: | | | | | |
| Home management | 74.36 | 75.92 | 84.76 | 85.80 | 94.12 |
| Home education | 1176.76 | 2207.4 | 3283.8 | 3063.84 | 3829.28 |
| Total | 1251.12 | 2283.32 | 3368.56 | 3149.64 | 3923.40 |
| HP(MHP+NMHP) | 3698.76 | 4712.76 | 5887.44 | 5053.88 | 5781.28 |

182

## Table 6.31
## Average weekly and annual dollar value of production
## By level of education of both spouses
## (Excluding the contributions of paid domestic help)

| Type of Household Activities | Per household weekly value ($) | | |
|---|---|---|---|
| | Both spouses in high education group | Both spouses in low education group | One spouse in high education and the other in low education group |
| MHP | 31.52 | 52.36 | 39.77 |
| NMHP: | | | |
| Home management | 1.98 | 1.55 | 1.90 |
| Home education | 76.84 | 42.42 | 76.03 |
| Total | 78.82 | 43.97 | 77.93 |
| HP(MHP+NMHP) | 110.34 | 96.33 | 117.70 |
| Type of Household Activities | Per household annual value ($) | | |
| | Both spouses in high education group | Both spouses in low education group | One spouses in high education and the other in low education group |
| MHP | 1639.04 | 2722.72 | 2068.04 |
| NMHP: | | | |
| Home management | 102.96 | 80.60 | 98.80 |
| Home education | 3995.68 | 2205.84 | 3953.56 |
| Total | 4098.64 | 2286.44 | 4052.36 |
| HP(MHP+NMHP) | 5737.68 | 5009.16 | 6120.40 |

183

The above results on the value of household production by level of educational attainment of wives are also consistent when the level of education for both spouses are taken into account. This is shown in Table 6.31. Families where both spouses are considered as in the low education group contributed the lowest value of household production ($96 per week, and $5009 per year). Again this lower value is attributed to the low dollar value assigned to the home education of children, which in turn, is the result of low hours devoted to home education (see Table 6.17 on time spent). Indeed, if the value of household production is based solely on MHP activities, then families, where both spouses are lowly educated would contribute the highest value (compare $52.36 to $31.52 for high education group and $39.77 for mixed education group). Families where one spouse is highly educated and the other low, contributed the highest value of household production (primarily through the value of home education).

**6.5    The GNP value of household production in 1986**

The aggregate or GNP value of household production in 1986 for all households in Singapore can now be derived using the average annual values of household production per household according to certain household characteristics. This is done by taking the reported annual dollar values of household production per household (see section 2) and multiplying this by the total number of households in the country having such a characteristic. One caveat is required. That because of the limitations of the secondary data on the number of households in Singapore having a particular household characteristic, in that, the latest *Report of the Household Expenditure Survey* published by the Department of Statistics is that of 1982-1983, we might expect that the numbers reported would have changed by 1986. Thus, the values reported below are more likely to represent conservative or at best minimum estimates for 1986.

Table 6.32 shows the value of household production according to the size of the household in Singapore (excluding the contributions made by paid domestic help). In 1986, the value of household production totalled $2.11 billion or about 5.38 percent of Singapore's GNP. This aggregate value is made up of about $1 billion worth of MHP activities and $1.11 billion worth of NMHP activities. Thus, while it appears that significantly less hours are devoted to NMHP activities by all households than MHP activities, the value-added of the former activities is much higher than the latter. The value-added of total household production for a family of size 2, 3 to 5, and 6 or more persons are about $93 million, $1.33 billion, and $688 million respectively.

Table 6.33 shows the value of household production according to the size of the household in Singapore when the contributions of paid domestic help are included. The total value of household production rose from $2.11 billion when domestic help is excluded to about $2.34 billion when they are included. As a percentage of GNP, this is about 5.97 percent. We might note that much of the increase in value comes from MHP activities, whilst the value of NMHP activities appears relatively stable. Again, this is because NMHP activities are seldom delegated to paid domestic help.

Strictly speaking, the dollar contributions made by paid domestic help to household production should be excluded in the calculation of the GNP value of household production since the wages received by paid domestic help are entered into the calculation of national income. Table 6.33 is included here only for comparative reasons with the results from Table 6.32.

Table 6.34 presents the total value of household production using the number of pre-primary school children in the family as the household characteristic. In 1986, this aggregate value of household production is about $1.53 billion or 4 percent of Singapore's GNP.[18]

Finally in Tables 6.35 and 6.36, the value of household production in Singapore is derived from the educational attainment of the wife, and of both spouses respectively. In the case of the former, the value of household production amounted to about $1.62 billion or 4.14 percent of Singapore's GNP. In the case of the latter, the value of household

---

[18]  We might note that the total value of household production derived from using the presence of pre-primary school children in the family as the household characteristic is less than the amount derived from using size of family as the household characteristic. This is because of the fewer reported total number of households in the population of the former than the latter. The total number of households with different numbers of pre-primary school children is reported as 353,400 households in the Household Expenditure Survey 1982-1983 published by the Department of Statistics (at page 250, Table I.109) whereas the same Report published 472,800 households for the total number of households of different family sizes. The difference of 119,400 households may be due to residual errors in the survey.

production is estimated to be $2.45 billion or 6.25 percent of Singapore's GNP.

Thus, irregardless of which household characteristic is used as the base from which the value of household production is derived, the results seemed to indicate that the value of household production ranges from 4-6 percent of Singapore's GNP. Even accounting for the caveat introduced at the beginning of this section, it might appear that the value of household production is rather low in comparison to estimates of 20 to 30 percent of most Western countries such as the United States and Canada.[19] However, this is hardly surprising since unlike the studies done in the West, in this study, no estimation is made on the contributions of paid domestic help to the GNP as their wages are already included in the calculation of national income.

Another major reason for the low estimates compared to other studies has to do with the methodology of estimation on quantity demanded of household production. This study excludes the time reported as spent on miscellaneous household activities and that where there are simultaneous activities, only the major task is to be reported. The study also utilises a valuation method that among other things, adjust for efficiency differences between a paid domestic help and the household member performing MHP activities. In the case of Singapore, this relative efficiency coefficient is shown to be favouring the paid domestic help in that hired-help is generally more efficient. Thus, in computing a value for MHP activities, a lower market replacement cost estimate reflecting efficiency differences is assigned to it.

In the next section, we compare the results obtained in this study using the efficiency-adjusted replacement cost method with the values of household production derived from the traditional methods, namely, the opportunity cost and the unadjusted replacement cost method using the same Singapore household data.

6.6     Comparing with estimates derived from the itemised replacement cost method and the opportunity cost approach using Singapore household data

The methodologies used to derive traditional itemised replacement cost estimates and opportunity cost estimates have been described earlier in

---

[19]     See Chapter 4.

Chapter 4. We now derive both sets of estimates and compare them to this study's efficiency-adjusted replacement cost results.

*Itemised replacement cost estimates*

The itemised replacement cost method assumes that all household production activities have market equivalents. The method then calls for a list of these market equivalents and their per hour wage rate. Multiplying the time spent in each of the household-market equivalent activity by its per hour wage rate and summing this over all household production activities yields the average per household value of household production. The average per household value of household production is then used to derive the total value of household production for Singapore by summing and multiplying over all households in Singapore.

Table 6.37 lists the household-market equivalent activities and derives the per hour wage rate for each of these activities. Using the weekly and annually reported time spent in household production by household activities and size of the household (Table 6.1 and Table 6.27 this Chapter), the time devoted to each household activity is then multiplied by its per hour wage rate derived in Table 6.37. This is shown in Table 6.40 (weekly value) and 6.39 (annual value). Finally, in Table 6.40, the total value of household production in Singapore in 1986 using the itemised replacement cost method is derived. This is obtained by taking the values reported in Table 6.39 and multiplying them by the number of households in Singapore of size 2, 3 to 5, and 6 or more persons.

In sum, the value of household production in Singapore in 1986 amounted to about $5.18 billion or 13 percent of Singapore's GNP using the itemised replacement cost method. This is more than twice the value of household production derived by the efficiency-adjusted replacement cost method (see Table 6.32). That this is so is not surprising since the itemised replacement cost method uses (1) specialist wages, and (2) does not adjust for efficiency differences between the market replacement and the household. While the estimated 13 percent household production value of Singapore's GNP still remains below itemised replacement cost estimates generated by Canadian and United States studies for example, it must be noted that the study here has excluded the dollar contributions of paid domestic help which are usually included in the other studies.

187

## Table 6.32
## Value of household production in Singapore by size of household in 1986[1]
(Excluding the contributions of paid domestic help)

| Types of Household Activities | Household Size | | | | Percentage of: | | |
|---|---|---|---|---|---|---|---|
| | 2 | 3-5 | 6 or more | Total | GNP[3] | GDP[4] | IGNP[5] |
| MHP | 45881.00 | 592406.89 | 363126.40 | 1001414.29 | 2.56 | 2.85 | 3.06 |
| NMHP: | | | | | | | |
| Home management | 3727.67 | 21167.54 | 13538.98 | 38434.19 | | | |
| Home education | 40442.69 | 713161.28 | 312100.88 | 1065704.85 | | | |
| Total | 44170.36 | 734328.82 | 325639.86 | 1104139.04 | 2.82 | 31.4 | 3.38 |
| HP(MHP+NMHP) | 93051.36 | 1326735.71 | 688766.26 | 2108553.33 | 5.38 | 5.99 | 6.44 |
| N[2] | 49.1 | 273.2 | 150.5 | 472.8 | | | |

1  In thousands of dollars

2  *Report on the Household Expenditure Survey 1982-1983*, Department of Statistics Publication, Page 229 Table 1.88  N = Number of households in the population.

3  GNP = $3918520000.  From *Economic Survey of Singapore 1986*, Ministry of Trade and Industry Publication, page ix.

4  GDP = $3518050000.  From *Economic Survey of Singapore 1986*, Ministry of Trade and Industry Publication, page ix.

5  IGNP = Indigenous GNP, defined as the GNP less the foreign shares in the GNP. It is different from the GNP in the sense that IGNP does not consider whether money earned in Singapore by foreigners get remitted abroad or not; as long as the money is earned by foreigners, it is subtracted.  The IGNP for 1986 is $3271500000.

**Table 6.33**

**Value of household production in Singapore by size of household in 1986[1]**

**(Including the contributions of paid domestic help)**

| Types of Household Activities | Household Size | | | | Percentage of: | | |
|---|---|---|---|---|---|---|---|
| | 2 | 3-5 | 6 or more | Total | GNP[3] | GDP[4] | IGNP[5] |
| MHP | 50732.08 | 717707.33 | 461734.00 | 1230173.41 | 3.14 | 35.0 | 3.76 |
| NMHP: | | | | | | | |
| Home management | 3191.50 | 24150.88 | 13617.98 | 40959.62 | | | |
| Home education | 40442.69 | 717161.28 | 312100.88 | 1069704.85 | | | |
| Total | 43634.19 | 741312.16 | 325718.12 | 1110664.47 | 2.83 | 3.16 | 3.39 |
| HP(MHP+NMHP) | 94366.27 | 1459019.49 | 787452.12 | 2340837.88 | 5.97 | 6.66 | 7.15 |
| N[2] | 49.1 | 273.2 | 150.5 | 472.8 | | | |

1   In thousands of dollars
2   *Report on the Household Expenditure Survey 1982-1983*, Department of Statistics Publication, Page 229 Table 1.88  N = Number of households in the population.
3   GNP = $39185200000. From *Economic Survey of Singapore 1986*, Ministry of Trade and Industry Publication, page ix.
4   GDP = $35180500000. From *Economic Survey of Singapore 1986*, Ministry of Trade and Industry Publication, page ix.
5   IGNP = Indigenous GNP, defined as the GNP less the foreign shares in the GNP. It is different from the GNP in the sense that IGNP does not consider whether money earned in Singapore by foreigners get remitted abroad or not; as long as the money is earned by foreigners, it is subtracted. The IGNP for 1986 is $32717500000.

## Table 6.34

### Value of household production in Singapore by number of pre-prima by school children in 1986[1]
### (Excluding the contributions of paid domestic help)

| Types of Household Activities | Number of Pre-Primary School Children (< = 6 years of Age) | | | | Percentage of: | | |
| --- | --- | --- | --- | --- | --- | --- | --- |
| | 0 | 1 | 2 or more | Total | GNP[3] | GDP[4] | IGNP[5] |
| MHP | 646827.32 | 331925.57 | 133778.89 | 1112531.78 | 2.84 | 3.16 | 3.40 |
| NMHP:<br>Home management<br>Home education | 18884.58<br>- | 6778.51<br>272425.30 | 2996.76<br>115357.63 | 28659.85<br>387782.93 | | | |
| Total | 18884.58 | 279203.81 | 118354.39 | 416442.78 | 1.06 | 1.19 | 1.27 |
| HP(MHP+NMHP) | 665711.9 | 611129.38 | 252138.28 | 1528974.56 | 3.90 | 4.35 | 4.67 |
| N[2] | 234.3 | 85.2 | 33.9 | 353.4 | | | |

1   In thousands of dollars
2   N = Number of households.  *Report on the Household Expenditure Survey 1982-1983*, Department of Statistics Publication, Page 250
    Table 1.109.
3   GNP = $39185200000.  See Note 3 Table 6.32.
4   GDP = $35180500000.  See Note 4 Table 6.32.
5   IGNP = $32717500000. See Note 5 Table 6.32.

## Table 6.35
### Value of household production in Singapore by the educational attainment of the wife in 1986[1]
(Excluding the contributions of paid domestic help)

| Types of Household Activities | Education Level | | | | | Percentage of: | | |
|---|---|---|---|---|---|---|---|---|
| | Primary School and Below[6] | Secondary School | Post Secondary | Tertiary | Total | GNP[3] | GDP[4] | IGNP[5] |
| MHP | 636215.09 | 149369.58 | 39989.04 | 20268.92 | 845842.63 | 2.16 | 2.40 | 2.59 |
| NMHP: Home management | 19604.03 | 5026.27 | 1901.80 | 1157.68 | 27689.78 | | | |
| Home education | 441463.67 | 194729.34 | 64340.64 | 47100.14 | 747633.79 | | | |
| Total | 461067.70 | 199755.61 | 66242.44 | 48257.82 | 775323.57 | 1.98 | 2.20 | 2.37 |
| HP(MHP+NMHP) | 1097282.79 | 349125.19 | 106231.48 | 68526.74 | 1621166.20 | 4.14 | 4.60 | 4.96 |
| N[2] | 260.90 | 593.2 | 21.0 | 12.3 | 353.4 | | | |

1   In thousands of dollars
2   N = Number of households. *Report on the Household Expenditure Survey 1982-1983*, Department of Statistics Publication, Page 253 Table 1.112.
3   GNP = $39185200000.  See Note 3 Table 6.32.
4   GDP = $35180500000.  See Note 4 Table 6.32.
5   IGNP = $32717500000. See Note 5 Table 6.32.
6   This category includes those wives without any formal education (ie. no schooling).

Table 6.36

**Value of household production in Singapore by the educational attainment of both husband and wife in 1986[1]**
**(Excluding the contributions of paid domestic help)**

| Types of Household Activities | Both spouses in high education group[6] | Both spouses in low Education group[7] | One spouse in high and the other in low education group[8] | Total | Percentage of: | | |
|---|---|---|---|---|---|---|---|
| | | | | | GNP[3] | GDP[4] | IGNP[5] |
| MHP | 76789.02 | 1043237.40 | 88491.43 | 1208517.85 | 3.08 | 3.44 | 3.69 |
| NMHP: | | | | | | | |
| Home management | 4823.68 | 30882.70 | 4227.65 | 39934.03 | | | |
| Home education | 18197.61 | 845189.65 | 169172.83 | 1201560.09 | | | |
| Total | 19021.29 | 876072.35 | 173400.48 | 1241494.12 | 3.17 | 3.45 | 3.71 |
| HP(MHP+NMHP) | 26810.31 | 1919309.75 | 261891.91 | 2450011.97 | 6.25 | 6.89 | 7.40 |
| N[2] | | | | | | | |

1  In thousands of dollars
2  N = Number of households.  *Report on the Household Expenditure Survey 1982-1983*, Department of Statistics Publication, Page 229
   Table 1.88.
3  GNP = $3918520000.  See Note 3 Table 6.32.
4  GDP = $3518050000.  See Note 4 Table 6.32.
5  IGNP = $32717500000. See Note 5 Table 6.32.
6  High Education Group = Tertiary + Post secondary school.
7  Low Education Group = No schooling + Primary + Secondary schooling
8  Mixed Education Group

## Table 6.37
## Market equivalent occupations and wage rates for each type of household activity

| Type of household activity | Market equivalent occupations | Monthly wage rate ($)[1] | Hourly wage rate ($)[2] |
|---|---|---|---|
| Cooking and meal preparation | General cooks | 969.37 | 5.51 |
| After meal clean-up | Kitchen assistants | 416.77 | 2.37 |
| Regular household cleaning | Cleaning service workers | 435.71 | 2.48 |
| Laundry and ironing | Laundresses | 566.84 | 3.22 |
| Shopping for groceries | Messengers | 596.94 | 3.39 |
| Gardening | Gardeners | 569.95 | 3.24 |
| Child care | Kindergarten or day care teachers | 360.00 | 2.05 |
| Home management | Managers of very small firms | 220 | 1.25 |
| Home education of children | Primary and Pre-Primary school teachers | 619.52 | 3.52 |

Source: Research and Statistics Department, Ministry of Labour, and the *Singapore Year book of Labour Statistics (1986).*

1. The monthly wage rate used here refers to the basic wage rate which includes NWC (National Wages Council) increases (if applies), and the employee's contribution to CPF (Central Provident Fund). Excluded are things like overtime pay, bonuses, fringe benefits etc. As the actual basic wage for each employee varies widely depending on the length of service, level of skills, and experience, only the starting wage-rate (ie. the lowest acceptable worker's wage) is used. Where there are two or more market equivalent occupations, a weighted average is used to derive the monthly basic wage rate.

2. The hourly wage rate is derived by dividing the monthly wage rate by 176 hours, where 44 hours represent the standard work week for most full-time occupations in Singapore.

## Table 6.38
## Average weekly value of household production per household
## by size of family using the itemised replacement cost method

| Type of household activity | Family Size | | | |
|---|---|---|---|---|
| | 2 | 3-5 | 6 or More | Any Size |
| **MHP:** | | | | |
| Cooking and meal preparation | 43.20 | 68.16 | 81.88 | 70.69 |
| After meal clean-up | 6.35 | 9.74 | 10.31 | 9.69 |
| Cleaning | 11.48 | 15.20 | 15.30 | 14.98 |
| Laundry and Ironing | 9.66 | 15.94 | 17.77 | 16.07 |
| Shopping for Groceries | 12.10 | 16.92 | 18.95 | 19.22 |
| Gardening | 2.49 | 3.60 | 4.50 | 3.79 |
| Child care | 6.89 | 40.47 | 33.31 | 36.14 |
| Total | 92.17 | 170.03 | 182.02 | 170.58 |
| **NMHP:** | | | | |
| Home management | 1.46 | 1.49 | 1.73 | 1.56 |
| Home education of children | 15.84 | 50.20 | 39.88 | 35.31 |
| Total | 17.30 | 51.69 | 41.61 | 36.87 |
| HP(MHP+NMHP) | 109.47 | 221.72 | 223.63 | 207.45 |

**Table 6.39**
**Average annual value of household production per household by size of family using the itemised replacement cost method (Excluding the contributions of paid domestic help)**

| Type of household activity | Family Size | | | |
|---|---|---|---|---|
| | 2 | 3-5 | 6 or More | Any Size |
| MHP: | | | | |
| Cooking and meal preparation | 2246.40 | 3544.32 | 4257.76 | 3675.88 |
| After meal clean-up | 330.20 | 506.48 | 536.12 | 503.88 |
| Cleaning | 596.96 | 790.40 | 795.60 | 778.9 |
| Laundry and Ironing | 502.32 | 828.88 | 924.04 | 835.64 |
| Shopping for Groceries | 629.20 | 879.84 | 985.40 | 895.44 |
| Gardening | 129.48 | 187.20 | 234.00 | 197.08 |
| Child care | 358.28 | 2104.44 | 1732.12 | 1879.28 |
| Total | 4792.84 | 8841.56 | 9465.04 | 8766.16 |
| NMHP: | | | | |
| Home management | 75.92 | 77.48 | 89.96 | 81.12 |
| Home education of children | 823.68 | 2610.40 | 2073.76 | 1836.12 |
| Total | 899.60 | 2687.88 | 2163.72 | 1917.24 |
| HP(MHP+NMHP) | 5692.44 | 11529.44 | 11628.76 | 10683.40 |

## Table 6.40
## Value of household production in Singapore using the itemised replacement cost method and by size of household in 1986[1]
### (Excluding the contributions of paid domestic help)

| Type of Household Activities | Household Size | | | Total | Percentage of: | | |
|---|---|---|---|---|---|---|---|
| | 2 | 3-5 | 6 or more | | GNP[3] | GDP[4] | IGNP[5] |
| MHP | 235328.44 | 2418166.66 | 1424488.52 | 4077983.62 | 10.41 | 11.59 | 12.46 |
| NMHP | 44170.36 | 735135.18 | 325639.86 | 1104945.40 | 2.82 | 3.14 | 3.38 |
| HP(MHP+NMHP) | 279498.8 | 3153301.84 | 1750128.38 | 5182929.02 | 13.23 | 14.73 | 15.84 |
| N[2] | 49.1 | 273.50 | 150.50 | 473.10 | | | |

1  In thousands of dollars.

2  N = Number of households.  See Note 2 Table 6.32.

3  GNP = $39185200000.  See Note 3 Table 6.32.

4  GDP = $35180500000.  See Note 4 Table 6.32.

5  IGNP = $32717500000. See Note 5 Table 6.32.

## Table 6.41
### Average net market income and education level by sex

| Sex | High Education[1] ($/Month) | Low Education[2] ($/Month) | Annual Gross Income ($/year) | Net of Tax ($/Year) | Hourly Wage Rate ($/Hour) |
|---|---|---|---|---|---|
| Male | 2820.5 | 767.5 | 33846 (H) | 29238.34 (H) | 12.78 (H) |
| | | | 9210 (L) | 8681.1 (L) | 3.79 (L) |
| Female | 1438 | 476 | 17256 (H) | 15740.16 (H) | 6.88 (H) |
| | | | 5712 (L) | 5462.16 (L) | 2.39 (L) |

Source: *Report on the Household Expenditure Survey 1982-1983*, Department of Statistics Publication page 57 Table 41, and Dhawn, S.L., *Singapore Income Tax*, Criterion Publishers 1985

1   The High Education Group is defined as those having completed tertiary and posted-secondary education. An average is taken.

2   The Low Education Group is defined as those having completed Secondary, Primary Education, and No Schooling. An average is taken.

### Table 6.42
### Average weekly and annual dollar value of household production by level of education of both spouses using the opportunity cost method[1]

| Per Household | Types of household activities | Both spouses in high education in low | Both spouses in low education Group | Husband in high education and wife education group | Wife in high education and husband in low |
|---|---|---|---|---|---|
| Weekly Value | MHP | 286.28 | 156.22 | 165.57 | 310.95 |
| | NMHP: Home management | 16.70 | 4.05 | 13.50 | 7.52 |
| | Home education | 194.49 | 62.69 | 128.83 | 125.65 |
| | HP(MHP+NMHP) | 497.47 | 222.96 | 307.90 | 444.12 |
| Annual Value ($) | MHP | 14886.56 | 8123.44 | 8609.64 | 16169.40 |
| | NMHP: Home management | 868.40 | 210.60 | 702.00 | 391.04 |
| | Home education | 10113.48 | 3259.88 | 6699.16 | 6533.80 |
| | HP(MHP+NMHP) | 25868.44 | 11593.92 | 16010.80 | 23094.24 |

[1] The values in the table are derived by the formula: Hours spent in household production x division of labour (%) x hourly wage rate for MHP activity, the husbands' share of production is 10.58 whilst the wives', 89.42. For home management, the husband's share is 62.50 and the wives', 37.50. For home education, the husbands' share is 34.40 and the wives', 65.60. For a detail discussion on division of labour, see my PhD dissertation, Quah, 1987.

## Table 6.43
## Value of household production in Singapore using the opportunity cost method and by the educational attainment of both husband and wife in 1986[1] (Excluding the contributions of paid domestic help)

| Types of household activities | Both spouses in high education group[6] | Both spouses in low education group[7] | One spouses in high and the other in low education group[8] | Total | Percentage of: | | |
|---|---|---|---|---|---|---|---|
| | | | | | GNP[3] | GDP[4] | IGNP[5] |
| MHP: | 697435.34 | 3112577.27 | 368406.50 (691888.63) | 4178419.11 (4501901.24) | 10.66 (11.49) | 11.88 (12.80) | 12.779 (13.76) |
| NMHP: Home management | 40684.54 | 80693.50 | 30038.58 (16732.00) | 151416.62 | | | |
| Home education | 473816.54 | 1249055.62 | 286657.06 (279581.30) | 2009529.22 | | | |
| Total | 514501.08 | 1329749.12 | 316695.64 (29631.33) | 2160945.84 (1873881.53) | 5.51 (4.78) | 6.14 (5.33) | 6.60 (5.73) |
| HP(MHP+NMHP) | 1211936.42 | 442326.39 | 685102.14 (98202.53) | 6339364.95 (6642465.34) | 16.17 (16.27) | 18.02 (18.13) | 19.37 (19.49) |
| N² | 46.85 | 383.16 | 42.79 | 472.80 | | | |

1   In thousands of dollars
2   N = Number of households.  Report on the Household Expenditure Survey 1982-1983, Department of Statistics Publication, page 229 Table 1.88
3   GNP = $39185200000.  See Note 3 Table 6.34
4   GDP = $35180500000.  See Note 4 Table 6.34
5   IGNP = $32717500000.  See Note 5 Table 6.34
6   See Note 6 Table 6.38
7   See Note 7 Table 6.38
8   See Note 8 Table 6.38.  The figures in the parenthesis denote the household characteristic where the wife is in the high education group and the husband is in the low education group.

*Opportunity cost estimates*

The second traditional method used to value household production is the opportunity cost method. The method uses the foregone wages of the homemaker as the value-added of time devoted to producing household services (see chapter 4). Strictly speaking, in applying the opportunity cost approach to valuing household production, it is the net return to paid employment that should be estimated. In deriving this net return to paid employment, marginal income tax-rates and work-related costs (such as transportation costs) must be subtracted from the homemaker's average gross market wages.[20] But because of the difficulty in obtaining an estimate of work-related costs, most researchers have used the average gross market wages or the after-tax market wages as measures of the opportunity cost.[21] The opportunity cost estimates derived below follow this tradition in that, estimates are based on after-tax market wages.

However, instead of using the forgone wages as measured by the average market wages of women as the opportunity cost, we use the average market wages net of taxes for women and men respectively, and according to their educational levels.

Table 6.41 presents some estimates of the per hour wage rate net of taxes for high and low educated males and females. Table 6.42 shows the weekly and annual dollar values of household production per household using the opportunity cost method. Finally, in Table 6.43, the total value of household production in Singapore is derived using the opportunity cost method.

The opportunity cost method reveals that the value of household production in Singapore is about $6-7 billion or 16-17 percent of Singapore's GNP. Compared to the efficiency-adjusted replacement cost method, this is more than twice (and nearly three times) the value of household production derived by the former method.

---

[20] Notice that this excludes considerations of the net-utility from working in the market and at home.

[21] There are also problems involving definition and identification of work-related costs. Further, it is not clear as to whether some of these costs are fixed or variable and when they are marginal.

*In sum*

The value of household production in Singapore in 1986 using the efficiency-adjusted replacement cost method appears to be about 4-6 percent of Singapore's GNP. This excludes the contributions made by paid domestic help in Singapore households. Opportunity cost and replacement cost methods tend to yield higher value estimates (more than twice), and if the values obtained by the efficiency-adjusted replacement cost method are even roughly correct to some order of magnitudes, they are thus suggestive of the extensive bias inherent in traditional procedures.

# 7 Conclusion

Major findings, use of household production estimates for public policies, and recommendations for future work

No one doubts that unpaid activities such as that of household production are valuable to households, and to a society, but despite this overt recognition, the dollar value of these contributions are neither treated as a separate entry nor included in a country's social accounts. This is because the value of household produced goods and services is difficult to measure and there are no market prices to indicate marginal values. It is these 'uncounted' household produced goods and services that have attracted and intrigued researchers over the years. But despite their valiant attempts especially given the paucity of data, some of the most basic conceptual, theoretical and methodological problems with regard to measurement and valuation of household production are still unresolved. Casual estimates from time to time provided by newspaper reports, popular magazines, and a host of other news media add on to much of the confusion, and indeed contributed to the lack of seriousness and attention on the part of official response.

A number of objectives were pursued in this study. They are:

1    To examine the major theoretical issues involved in valuing household production; specifically on the definition, quantification, method of accounting for joint-production activities, and on the appropriate method for valuing household production which is consistent with economic theory;

2    To provide a general framework for the valuation of non-market household production consistent with the way market production of goods and services are valued; and

202

3	To measure the quantity and economic value of household production in Singapore for social accounting purposes.

This study notes that a social accounting of the value of non-market household production is necessary for a number of reasons. Among others, it is believed that the research output will be of benefit in policy planning and decision making in a variety of ways, and are necessary for the following purposes:

1	To measure with greater accuracy a society's total economic production. Most of the early studies on household production were concerned with the limitations imposed by the national accountants as to which activities are to be considered as economically productive and hence to be included in a country's national accounts, and which are not -- the so-called 'production boundary problem.'

These studies have contended that from a conceptual and theoretical standpoint, non-market productive activities of which household production is a large component, should be taken into account when assessing total flows of national income generated during a given period of time in a given society. Non-market household production cannot and should not be considered as a pure hobby or leisure but rather a flow of productive services capable of being marketed, and labour efforts and time allocated to it can often be channelled into equivalent market activities where they are accounted only because they are renumerated.

When considering the usefulness of these efforts to expand the national accounts into areas of non-marketed but economically productive activities, it is worth recalling the words of Ruggles and Ruggles (1970) that the primary objective of the social accounts is to provide an 'information framework,' and that this framework extends more than the final set of accounts. Developing a set of ancillary data helps to assure completeness in coverage, and more importantly assures definitional consistency with other economic data.

2	The evaluation of the dollar value of non-market household production provides information as to the implicit price at which households value the use of their time in the provision of household services, and the consumption of household output vis a vis market services and the consumption of market output.

Knowledge of this information will in turn provide useful analysis on the changes in the distribution of labour between the market sector and the non-market sector or specifically on the labour force participation rates of women in the economy.

Economists have long been interested in the factors affecting the labour force participation of women relying much on the Mincer schooling model and Becker's opportunity cost of time model. In more recent models on labour force participation of women (Gronau, 1979, for example), non-market activities such as that of household production are introduced where the value of household production was derived using the ages of children as the determining characteristic of the household. Since the value of household production can now be derived directly using a list of household characteristics significant to its variation, labour supply functions can be estimated with greater accuracy. The interest in these evaluations it seems then lies in the indications they give on the consumption of a scarce resource, namely labour time by home activities and by market activities.

While the use of household production estimates may improve labour supply forecasts, it is less valid for long-run time series than for cross-section or shorter time series. This is because of the lack of complete time series data on time use in household production and consequently on valuation estimates. It is thus desirable to have more and regular information on household production estimates obtained through official bodies rather than the 'once-in-a-while' efforts of the past.

3    To identify, analyse and forecast which market sectors are (or will be) expanding or contracting as a result of shifts of market activities to the household, or vice-versa. While time-budget studies have revealed that the labour force participation of women has increased over several decades, it has not resulted in a significant reduction in the time devoted to household production. But time-budget studies also show that there has been a significant reallocation of time among the various types of household production activities. They have reported that there has been a reduction in time devoted to cooking, washing, cleaning and other physical domestic chores, but an increase in time spent on child care, and on the home education of children.

The result of this revealed preferred reallocation of time among household activities has been a substantial rise in the commercialisation of household work. This, in turn, can be attributed to rising family incomes. Examples on commercialised housework abound: fast food chains, cleaning services, launderettes, etc. In the case of Singapore, this commercialisation of household production has taken the form of widespread maid services, launderettes, day care centres, and hawker centres. Useful insights may be gained in labour supply analysis of women by considering for example, what percentage of women stays at home because of the absence of services which act as substitutes for household activities, such as child care services etc.

The macroeconomic impact of this commercialisation of household production is the obvious creation of employment and income opportunities. Changes in technology or modes of household production are likely since market equivalent services would take the form of mass production, and hence one can expect a secondary growth in technology-related industries. If such technology changes market output or efficiency, and can be adapted to small scale production, then, this, in turn, will change family household production technology.

While some market sectors will expand, others, for example, the launderette industry may decline following a rise in the use and preference for home washing machines. Another example is the widespread and increasing use of home micro-wave ovens and its dampening effect on the spread of fast food stores. Thus, the growth potential of this market equivalent household production service sector has its limits; that limit being future demand for its market product. Household production research and estimates can therefore prove useful to business observers and policy planners in delineating these limits to growth. In other words, one can predict with some accuracy which industries can be expected to expand or contract, and their maximum growth potential.

4    Related to the macroeconomic implications arising from the commercialisation of household production and specific to the Singapore economy is the 'foreign maids issue.' Success in current Government policies aimed at inducing women to join the market work force is one of the major contributing factors in the rapid rise and profitability of commercialised housework services. Because there is a shortage in the supply of local Singaporean maids (and consequently a higher price for such maids), the employment of foreign maids becomes attractive as a relatively 'cheaper'

alternative. While no doubt filling up the household production gap vacated by market-seeking homemakers, the influx of foreign maids and the increasing overdependence of Singaporean households on these maids have led to rising social problems. What factors determine the hiring of foreign maids are thus necessary areas of research in formulating policies. In this study, it was shown that the value of household production was an important endogenous factor.

5    To measure economic growth more accurately. The exclusion of household production estimates from official national accounts has been shown by Weinrobe (1974) to result in an upward bias in GNP growth estimates. This is because of the increasing rate of women entering the labour force with a consequent decrease in household production. Total real output as such should grow less than the measured market output. Thus, fiscal and monetary policies aimed at decreasing inflation and/or unemployment could be more effective if household production estimates are considered.

6    As part of an important social indicator of economic well-being or welfare. Just as production of food for one's own consumption contributes to economic welfare so should it be the case for the social accounting of household production. Together with other social indicators like health, education, and status of the environment, information on household production estimates tell us something about the composition, structure, or functioning of a society, and measure changes in these aspects over time. To the extent that these changes are positive over time, say, more educated people, better health facilities and services, higher household production values, ceteris paribus these social indicators may be taken to reflect improvements in economic well-being of that society. The GNP statistic is but one of the social indicators.

7    To provide a better measure of family income distribution. The validity of any study on income distribution depends on at least five of the following factors: (i) the concept of family income or its definition, (ii) the time period, (iii) the definition of income expenditure, (iv) the measure of income distribution, and (v) statistical errors in measurement. Thus, if a wide definition of family income is adopted such that it includes both money income

and non-market income, and since household production activities are the main component of the latter, it is clearly necessary then to include an estimate of the value of household production.

As far as this author is aware, there have been no attempts nor any studies done to evaluate the distribution of total family income (which includes household production estimates) as compared to family money income. Perhaps this is an area for future research.

8    Another public policy application of household production estimates is the possible use of discriminatory taxation to affect labour force participation decisions from non-market to market and vice-versa. Currently household production activities while clearly contributing to total household income, are untaxed. The zero-tax on household production is thus an inducement for some people to stay out of the market sector. Household production estimates by age, household size, education of spouses, and other household characteristics, would surely be useful guidelines for policy makers to influence household members' decisions with regard to the use of their time in market work, housework and leisure. This, may in turn, influence productivity measures in market and home work.

The above represents at least 8 separate areas of potential uses for household production estimates in the conduct of public policy formulations. Two further but controversial applications of household production estimates on public policy which have received considerable attention in recent years are the proposals for salaried homemakers, and their pensionability.

Because most household production is undertaken by women, the proposed scheme of wage payments for homemakers has been popularly debated as a 'wages for housewives' context. There are three major arguments for the provision of a wage to full-time homemakers: (i) the economic dependence and insecurity of homemakers because they are not paid for doing household work; (ii) the responsibility of society to reward homemakers to maintain the adult-working labour force, by ensuring its continued future supply from the production of children at home; and (iii) as a symbolic recognition of the real work done by homemakers so that the wages paid would serve as an imputed price for use in GNP measurement, matrimonial property settlements, and tort compensation.

However, a more pertinent question is to ask who is going to make this wage payment? Presumably this question can be answered by asking another question, and that is, who derives the benefit from the homemaker's services? Is it the husband, wife, family, the employer, and/or society?

The argument that society ought to pay wages for homemakers rests on the assertion that there exists a positive social externality to raising children at home than say leaving them at child care centres during the working day. This has not been proven and remains to be shown. Furthermore, the money wage assigned to homemaking should measure only the positive externality generated on society, and should not include any benefits captured by the family or its individual members. If these two assertions or conditions are met, then arguably the payment by society to homemakers can be justified since a positive social externality ought to be encouraged and rewarded.

The suggestion that the husband should make the wage payment is one that is fraught with the most difficulty. This is because a marriage is both an economic and social relationship. In other words, household members or spouses share not just an economic relationship (as in the New Home Economics tradition) but also as social beings having a social relationship and interaction within a family unit. This is further complicated by the difficulty of dividing the household services performed by the homemaker for the husband, other family members, and for herself. Additionally, it has been shown by Bergman (1981) that the homemaker housewife is already in receipt of a payment in-kind from her husband equivalent to about 50 percent of her husband's income.

The third argument that it is the husband's employer who should pay for the homemaker's services is theoretically incorrect. This argument is based on the belief that the household production (inclusive of child care) performed by the homemaker housewife allows the husband to be available for market employment and because he need not worry about the household work and child care at home, he can concentrate on his market work, with the result that his productivity and efficiency is said to increase.

However, the theoretical flaw of this argument is that, if the labour market is assumed to function competitively the wage received by the market worker should equal his marginal product inclusive of the wage effect. Thus, the wage paid to the husband by his employer would already have incorporated the indirect services of the homemaker housewife.

In sum, the wage payment to homemakers suggestion appears to be very problematic in application and given the lack of economic justification for such a payment other than in the case of positive externalities conferred on society (and even this, depends on the establishment of such benefits), it is perhaps desirable not to pursue this wages-for-homemakers idea to promote equity and/or establish an imputed price for household services. Further, as Windschuttle (1972) has noted that if a salary is paid to all homemakers in Australia, say in the region of about A$20 per week, this would amount to nearly half of Australia's social security budget!

The other proposal calling for the pensionability of homemakers appears to have some theoretical validity but just as the wages-for-homemakers suggestion, this proposal severely lacks practical application. Just as a market worker has the privilege to participate in some pension plans or social security systems, so may it be argued for the non-market worker engaged in economic household production. The immediate problem is of course the extreme difficulty of determining an income base of the household production contributions for the different and varied households. Thus, this would be an area where reliable and detailed household production estimates could aid in resolving much of the problem. But again the problem of who actually pays for this income-in-kind contribution of the homemakers exists and as such the problems associated with the wages-for-homemakers proposal resurface.

On the basis of the above investigation, it would appear that a social accounting of household production and its estimates can very evidently be used in a number of public policy formulations. Basic to the measurement and valuation of household production however, requires an official recording of non-market work time. Just as the consumer price index which records price changes, so should information on time accounts on a national scale be set up, and perhaps meaningfully referred to as the 'time value index.' To answer many of the questions which policy makers may have on the evaluation or proposition of certain programmes, the need or use of a time value index is strongly evident.

Estimates on household production should become a standard component of an official statistical series, whether these estimates be used to augment GNP measures and/or as a social indicator of non-market but economic activity. Data on household production should be available at regular intervals, perhaps every 5 years. If and when national time accounts are established, all kinds of sub-populations can be analysed for many official purposes.

The methods of collecting time use data need to be refined for some degree of accuracy. If collection of large census-like quantities of data is not feasible, then sample stratification by family characteristics may be of prime importance for valuing household production. Household production in consuming approximately at least equal to market work time deserves more attention and research than it has been given until now. If meaningful decisions are to be taken in the economic, social and manpower fields, economic measurements and evaluation of non-market household production have to be taken into account.

To recapitulate; the major findings and conclusions reached in this study include the following:

1      The term 'household production' must be defined unambiguously if any meaningful empirical data is to be collected for study. The definition of household production would, in turn, greatly depend on the specific objective of measurement and valuation of household production. Thus, for example, if the objective of valuation is for national income accounting purposes, then to be consistent with the valuation of market-based activities, the definition of household production should be one that is capable of market valuation. On the other hand, if the valuation is for own-family welfare considerations, the definition of household production would have to be broader and should include such things as conjugal relations, love and care, and family interaction.

For the purposes of this study, household production has been defined as consisting of market replaceable household production (MHP) -- those unpaid home activities which could be done to prescribed specifications and to the benefit of the household by someone outside of the household, usually through purchased domestic help services -- and, near market replaceable household production (NMHP) which are those unpaid home activities performed by family members in addition to market replaceable household production activities, and which do not find easy market replacements in the form of paid domestic help but could conceivably be done by appropriate specialised help. By defining household production in this way, we clearly exclude those activities in the home which are personal and hence non-marketable (such as love, care, and conjugal relations). The latter activities also lie outside the perimeter of possible imputations of non-market goods and services consistent with the social income accounting framework (see Chapter 3).

210

2    There are immediate advantages in using time as a unit of measurement and valuation for household production. First, time is additive, expressible in different units (that is, divisible) whereby allowing easier data collection on household contributions. Second, it is commensurate with market employment remuneration in that just as the market worker is most often paid in dollars per unit time, so can we derive the dollars per unit time for the home worker. The latter advantage in turn allows an income approach to valuing the work of family members engaged in household production. But more importantly, what is missing from the national accounts is not so much as the contribution of capital inputs in household production but rather the value-added of time used up in the production of household output. While market purchased goods are important inputs, they are already captured in the GNP the moment they leave the factory. Adding them to household production valuation for a second time would result in double-counting.

3    A more accurate method of accounting for time use in households where joint production activities are involved is to abscribe the time entirely to the major task. In other words, time is recorded only for the major household task performed and not other minor activities. Thus, whether the time use involves two or three simultaneous activities, it is left to the reporting unit to decide which is the major activity. This eliminates the need for researchers to weigh the relative importance of each of the tasks performed.

Recording for time use only for the major task involved appears to be a better solution since it provides a means of checking possible exaggerations of time devoted to household production. Also, the method works within the framework of a 168-hour week constraint, so that a micro-allocation of time for market work, house work and leisure, can be derived and the implications of such an allocation of time for the different households easily drawn.

4    The appropriate method of valuation of household production should depend critically on the purpose of the valuation: whether it is for national income accounting, matrimonial property settlements, or valuation for compensation. This question of

purpose has been seriously neglected by most researchers. Thus, depending on the purpose of valuation, the appropriate measure would involve marginal (as in national income accounting), total or net valuation (as in compensation for welfare loss through wrongful death or injury). It was suggested here that the theoretically correct method for valuing household production for social accounting purposes is the sum of the values of market replaceable household production (MHP) and near-market replaceable household production (NMHP). In deriving a value for MHP, a general model was constructed using utility arguments such as market and home production and consumption. The results from the model showed that while using the market replacement wage rate may serve as a useful approximation of the value of MHP, a more accurate imputation would depend on whether the household is as efficient as the hired market replacement. Thus, the model confirms the often-heard criticism of the replacement cost method which uses the market wage rates of hired-help will lead to over or under-estimation of the value of household production depending on the magnitude and direction of efficiency differences between the hired-help and the family-household.

The correct imputation for valuing MHP is thus the efficiency-adjusted replacement wage rate of a hired domestic help and not just the replacement wage rate. For the NMHP activities, defined as comprising of home management and the home education of children, the imputation of value took the form of using the market wage rates of small firm managers and that of kindergarten and primary school teachers respectively.

It was argued that because NMHP activities are not normally included in the hired services of a paid domestic help (nor are they appropriate for these kinds of work), it would be incorrect to impute a value for NMHP by taking the average wage rates of domestic help just as one does for valuing MHP. It was noted that past studies often did not make such distinctions. While admittedly a crude way of valuing NMHP, the method at least gives us a more accurate picture of the kind of labour services one normally finds in households whose activities have near market analogues and therefore capable of pecuniary imputations. Because the GNP measure is first and foremost a market based evaluation, the use

212

of market analogues to valuing household production is theoretically more appealing.

5    The average time used in household production (exclusive of paid domestic help's time) by families of size 2, 3-5, and 6 or more members is about 32 hours, 69 hours, and 67 hours per week respectively.

Time spent in MHP activities tend to vary systematically in an upward direction in that the larger the size of the household, the more hours were spent. This is true for both households with paid domestic help, and for those without paid domestic help.

Excluding NMHP activities, it was found that the most time-consuming activities in descending order are child-care, cooking and meal preparation, regular housecare or cleaning, grocery shopping, laundry work and ironing, after meal clean-up, and gardening or outdoor work. This ranking of activities by time devoted to household production is the same whether NMHP activities are included or excluded except that it was found that the time spent in the home education of children would rank about the same as the time spent in cooking and meal preparation but less than the time devoted to child care. The time spent in home management tends to be relatively stable and only about less than 2 hours a week across all household sizes. Thus, it appears that home management is a relatively insignificant activity despite the often a priori assertion (this, and other studies) of its importance.

6    Women spouses devoted significantly more time to household production whether the household hires a maid or not. It was found that men spouses spent on average 9.83 hours a week on household production (about a 15 percent share) while women spouses spent on average 51.26 hours a week (about a 77 percent share). The disparity in time spent in household production between husbands and wives is larger in the case of MHP activities than in NMHP activities, implying that men have less of a comparative disadvantage at NMHP activities than women. Also, NMHP activities need not be done on a daily basis, whereas MHP activities are more regular; for example, meals have to be cooked, and children have to be looked after. The rest of the household

members, namely, the other adult members (relatives living-in) and children of the household spent a small and insignificant time in household production.

7    The following explanatory variables appeared significant in explaining variations in MHP: family Size (positive effect), age of the youngest child (negative effect), the number of times a household eats out a week (negative effect), and the number of primary school-going children (positive effect). Some differences were observed between households with maids and those without maids in that for those households without maids the following variables were found to be significant in addition to the above variables: household income (negative effect), location of households, and wife's employment status. It would seem that a possible explanation for this difference has been the observed lack of sample variability in these three variables in the case of households with maids than it is for households without maids.

8    The following explanatory variables appeared significant in explaining variations in NMHP: age of the youngest child (negative effect), number of primary school-going children (positive effect), husband's educational level, and wife's educational level. This confirms a number of a priori assertions (see Chapter 6) in this as well as findings from other studies (Walker; 1976 for example). Some differences were also observed between households with maids and those without maids, namely the type of dwelling (positive effect), household income (positive effect), and the number of times a household eats out (negative effect) were significant enough to explain variations in NMHP for households without maids but not for households with maids. Again the lack of sample variability in the case of households with maids may have accounted for this. However, the results of a Chow test on the equality of coefficients showed that all 3 demand for household production functions were not significantly different for the two groups (with and without maids) of households (see Chapter 6).

It appears that household equipment such as labour saving appliances were not significant in explaining variations in MHP and NMHP. This result confirms the findings of pervious studies. It appears also that the best regression model on the demand for household production is that which uses MHP as the dependent variable. This may be due to the lack of variation for all households (with and without maids) in time spent in NMHP activities, thus the low $R^2$ reported in the NMHP and HP(MHP + NMHP equations).

9    The average weekly and annual dollar values of household production respectively amount to $35 and $1834 for a family of size 2, $93 and $4856 for a family of size 3-5, and $87 and $4576 for a family of size 6 or more persons. The weekly and annual dollar values of household production for households without pre-primary school children amounted to about $54 and $2841 respectively. The presence of one pre-primary child adds about $83 and $4331 to the weekly and annual dollar value of household production respectively. The value of household production rises with more pre-primary school children in the family. The value of household production was lowest in families where the wife had no schooling, and this was attributed to the lower dollar values assigned to the home education of children, which in turn was the result of the reported low hours devoted to home education by these families.

For households without maids, the value of household production is about 10 percent of family money income on average. For households with maids, the value of household production is about 3 percent of family money income. This difference may be largely explained in that for those households with maids, and hence the relatively well-off tend to participate in more market-based production rather than household production so that the latter activity forms a smaller proportion of money income.

10    The magnitude of the value of household production according to the size of the household in Singapore is, for 1986, about $2 billion compared to the GNP for that year of about $39 billion.

In percentage terms, this is about 5 percent of Singapore's market GNP. This aggregate value is made up of about $1 billion (2.56%) worth of MHP activities, and $1.11 billion (2.82%) worth of NMHP activities. Thus, while it appears that significantly less hours are devoted to NMHP activities by all households than MHP activities, the value-added of the former activities is much higher than the latter. In comparison to other national estimates, particularly those from North America, the value of household production derived for Singapore appears to be proportionately lower. Whereas the North American studies show household production to average below 20 to 30 percent of their respective GNPs, the Singapore figures are only 5 to 6 percent. The difference lies in three factors: first, unlike the studies done in the West, estimates of household production in this study do not include the dollar contribution of paid domestic help to the GNP as their wages are already included in the calculation of national income; second, the methodology of estimation on the quantity of household production demanded is different from that utilised in most of other studies. For example, the present study excludes the time spent on miscellaneous household production activities (thus avoiding inclinations to exaggerate the time spent), and that where there are simultaneous activities, only the major task is reported; and third, this study utilises a valuation method that among other things, adjusts for efficiency differences between a paid domestic help and the household member performing MHP activities. In the case of Singapore, this relative efficiency coefficient is shown to be favouring the hired-help over the household member in that the former is, on average, generally more efficient than the latter. Previous studies assumed equal efficiency and hence by multiplying the replacement wage rates by the number of hours spent in household production would lead to higher value estimates.

It might be interesting to note therefore that when we recalculate the value of household production using the traditional replacement cost and opportunity cost approaches for the same Singapore household data, the values rose from 5-6 percent of GNP (using the efficiency-adjusted replacement cost method) to about 13-17 percent. Clearly then the magnitudes become more comparable to that of the North American studies. Thus, opportunity costs and replacement cost methods tend to yield higher value estimates (at least more than twice); and if the values obtained by the efficiency-adjusted replacement cost method are even roughly correct to some order of magnitudes, they are thus suggestive of the extensive bias inherent in traditional evaluation procedures.

216

*Limitations of study*

While the results reported in this study add valuable insights into the household economy, they must be regarded as tentative and require further verification. The measurement and evaluation approaches need to be more fully refined; specifically on household efficiency measures, better time recording, and on the evaluation of near-market replaceable household production activities (or NMHP). Several reservations may have been too easily brushed aside while others may have been completely overlooked. Other shortcomings may include hypothetical bias -- the result of not confronting the individual with a real situation but rather an assumed situation -- and, strategic bias where individuals may be induced to misstate their own preferences or responses in an attempt to influence the average statistic to as close as possible to their own.

But note that the existence of both a hypothetical bias and a strategic bias may in effect be incompatible, and may cancel each other out since the former implies that people are not likely to take these studies seriously because of the use of hypothetical situations while the latter implies that people are likely to be serious enough as to attempt to misrepresent their true preferences in an effort to influence the final result or outcome of the study.

In any case, it is appropriate to minimise hypothetical bias by paying close attention to the design of survey questionnaires and contingent questions. This requires for example, making the contingent setting believable to the respondent and that the researcher (or interviewer) must ensure that the respondents are fully aware and have understood all the ramifications of the proposed change. Other sources of bias, such as sampling bias, non-respondent bias, information bias and interviewer bias may also be present, but since these forms of biases are not particular to this study, as they are just as likely to occur in any other studies using survey techniques, they are not discussed here.

Despite some of these shortcomings, the present study has several unique features which distinguish it from past studies. These features include:

1   It explores the major theoretical and methodological issues involved in measuring and valuing household production, their persistent problems, and suggests ways of resolving them. Past studies either tend to ignore them or paid little attention to them. Instead, most studies were either very empirically based or

mathematically based.

2 The study suggests a new way of evaluating household production for social accounting purposes. The method suggested has been shown to be consistent with social accounting practices. This is done through a mathematical model which formally shows that for social accounting purposes, it is the efficiency-adjusted replacement cost method that appears appropriate.

Further, unlike existing formulations on household production which tend to concentrate on the dichotomy between market work and homework and the household's optimal allocation of time in these two sectors, the model specifically includes the services of domestic help to substitute for some of the household production activities currently undertaken by household members. The model is formulated in such a way that work by hired-help gives utility by increasing household consumption at home. The model also allows for efficiency or productivity differences in household production between a paid domestic help who works only in the home and a household member who works both at home and in the market.

3 It is the first comprehensive and systematic study on household production in Singapore. There have been the occasional newspaper surveys but other than that, no other studies exist. It thus provides a source of 'rich' information for future researchers.

4 The results from the household survey and its analyses reinforce some of the basic implications of the theory of household production, and reaffirm some of the results found in other empirical studies (cited in Chapter 4).

5 In Chapter 4 of this study, a comprehensive and up-to-date survey of major empirical studies on household production in the United States, Europe and Asia, the methods used in these studies, and their results were analysed and reported. The survey should benefit future researchers interested in empirical-based evaluations of household production in that it informs them what has been accomplished thus far. The studies reported are chronologically dated, and listed by country origins.

218

A lot still has to be done. National time accounts collected on households on an ongoing basis will be essential if the measurement and evaluation of household production become a serious national goal. This study, for example is restricted to consider only cross-sectional analysis of household production estimates. It says nothing on intertemporal changes in these estimates in Singapore simply because there are neither official nor unofficial studies on household production in Singapore before. Thus, the study was forced to exclude a time-series analysis of household production estimates.

The United Nations System of National Accounts (SNA) is currently under review. Among many of the topics, the United Nations Statistical Office has declared that it "will examine the possibilities and obstacles of further expanding the coverage of activities beyond the present SNA limits .... The study should build further on present research efforts" (United Nations, Economic and Social Council, 1984, pp. 17-18).

Further, at the most recent meeting of the Asean Women's Programme held in Kuala Lumpur, Malaysia (20-23 January 1986), the Philippine delegation comprising of officials from the National Accounts Staff of the National Economic and Development Authority proposed a collaborative effort in devising some form of measurement or indicator that could estimate the national value of household production. Both Malaysia and Indonesia immediately supported the Philippine proposal while the other member countries of Asean have agreed to consult their respective government agencies. The moment then seems appropriate for calling on an international collaboration for defining measurement and evaluation guidelines on household production.

While the value of household production is indeed useful for a number of purposes, its usefulness by no means requires that it be incorporated into the GNP. Indeed a meaningful procedure maybe to have a separate reporting of both GNP and household production estimates. Research is also needed to determine exactly how household production estimates can be incorporated into the GNP for those who argue in its favour. No attempt is made to expand upon this question in the present study. This, and removing some of the limitations reported above in this study would perhaps constitute a direction for future research but for now some preliminary work and/or foundation has been laid.

# Appendix
# The household survey

NATIONAL UNIVERSITY OF SINGAPORE

Kent Ridge
Singapore 0511

Department of Economics and Statistics

## Household Economics Survey

### Definition and Classification of Household Activities

## HOUSEHOLD ACTIVITIES

1.   Cooking and Meal Preparation

Preparation and cooking of food for all regular meals e.g. breakfast, lunch and evening meals

Preparation of regular drinks at all meals and for drinking throughout the day Setting the table

Serving the food

Preparation of snacks and packed lunches for household members to take to work and to school

Exclude food and drink preparation for special occasions e.g. parties, birthday and marriage anniversary celebrations and other irregular functions

2.   After Meal Clean-up

After meal care of dishes, left-overs, kitchen equipment and utensils, cleaning up of table and refuse/garbage removal

3.   Cleaning or Regular House Care

Daily cleaning activities e.g. mopping the floor, dusting, sweeping, vacuuming, tidying up the living room and other rooms, washing basins, cleaning of washrooms

Making beds

Washing windows and window ledge

4.   Laundry including ironing

Washing clothes at home or at a laundromat e.g. hand washing and/or machine washing

Collecting and sorting soiled clothes for washing

Pretreating with chemicals and other preparation of clothes for washing

Loading and unloading clothes from the washing machine and/or dryer

Exclude time taken by machine

Hanging clothes on line or pole for drying

Preparation of clothes for ironing

Actual ironing and pressing of clothes

Folding and storing ironed clothes

5.    Supervision of household tasks, accounts or matters of paper work

Settling bills and making cheque payments

Figuring out how much money is available (budget planning)

Recording and filing of receipts and expenditures

Making bank deposits

Planning menus for meals at home

Making out grocery lists and market/shopping items

Supervision of paid domestic help if any

6. Shopping for groceries and other weekly required items

> Actual time spent in shopping for groceries and other weekly required items whether in person, telephone, mail or home sales and delivery for household members

> Exclude shopping for personal items and window shopping or browsing (and for which no item is bought)

7. Outdoor Work/Gardening

> Daily or weekly care of garden areas, indoor house plants, washing of outdoor walkway, care of garbage and trash and all forms of maintenance of the house yard

> Washing and polishing of cars

8. Child Caring

> All activities related to the physical care of children e.g. bathing, feeding and dressing

> Preparation of milk bottles

> Changing of diapers

> Taking children to school and back

> Bedside care

> Watching attentively on a child's physical activity at home

> Playing with children
> Exclude the time of the mere physical presence of an adult member with the child (with no active participation of child caring by the adult(s).

****************

# Bibliography

Adler, H.J. and Hawrylyshyn, O., (1978), "Estimates of the Value of Household Work, Canada 1961 and 1971", *The Review of Income and Wealth*, Series 24, No. 4, pp. 333-355.

Alexander, J., (1981), "The Political Economy of Household Work", (Unpublished), Regina: University of Regina.

Anker, R., (1983), "Female Labour Force Participation in Developing Countries: A Critique of Current Definitions and Data Collection Methods", *International Labour Review*, Vol. 122, No. 6, pp. 709-723.

Becker, G.S., (1965), "A Theory of the Allocation of Time", *Economic Journal*, Vol. 75, pp. 493-517.

Becker, G.S., (1976), *The Economic Approach to Human Behavior*, Chicago: University of Chicago Press.

Becker, G.S., (1981), *A Treatise on the Family*, Harvard University Press.

Bell, E.B. and Taub, A.J., (1982), "The Value of Household Services", *American Journal of Economics and Sociology*, Vol. 14, No. 2, pp. 214-218.

Bender, M., (1974), "How Much is a Housewife Worth?", *McCall's New York*, May, pp 56.

Bergmann, B.R., (1981), "The Economic Risks of Being a Housewife", *American Economic Review*, May, pp. 81-86.

Berk, R.A., (1980), "The New Home Economics: An Agenda for Socio-logical Research", in S.F. Berk (ed.): *Women and Household Labour*, Sage Publ., pp. 113-148.

Berk, R.A. and Berk, S.F., (1983), "Supply-Side Sociology of the Family: The Challenge of the New Home Economics", *Annual Review of Sociology*, Vol. 9, pp. 375-395.

Beutler, I.F. and Owen, A.J., (1980), "New Perspectives on Home Production: A Conceptual View", in C. Hefferan (ed.), *The Household As Producer: A Look Beyond the Market*, American Home Economics

Association, pp. 15-30.

Beutler, I.F. and Owen, A.J., (1980), "A Home Production Activity Model", *Home Economics Research Journal*, September, pp. 16-26.

Bishop, R.C. and Heberlein, T.A., (1979), "Measuring Values of Extramarket Goods: Are Indirect Measures Biased?" *American Journal of Agricultural Economics*, Vol. 61, pp. 926-930.

Bivens, G.E. and Volker, C.B.; (1986), "A Value-Added Approach to Household Production: The Special Case of Meal Preparation", *Journal of Consumer Research*, Vol. 13, pp. 272-279.

Bloch, F.E., (1975), "The Allocation of Time to Household Work", Working Paper No. 64, Industrial Relations Section, Princeton University, pp. 1-12.

Blundell, R., (1981), "Taxation, Demographic Effects and Rationing in a Random Preference Model of Female Labour Supply", Discussion Paper No. 81-13, Department of Economics, The University of British Columbia, pp. 1-21.

Bockstael, N.E. and McConnell, K.E., (1983), "Welfare Measurement in the Household Production Framework", *The American Economic Review*, Vol. 73, pp. 806-814.

Boskin, M.J., (1974), "The Effects of Government Expenditures and Taxes on Female Labour", *American Economic Association - Women At Work*, Vol. 64, No. 2, 251-256.

Brody, W.H., (1975), *Economic Value of a Housewife*, Research and Statistics Note No. 9, DHEW Pub. No. SSA 75-11701, Washington, D.C., United States Department of Health, Education and Welfare.

Brookshire, D.S. and Crocker, T.D., (1981), "The Advantages of Contingent Valuation Methods for Benefit-Cost Analysis", *Public Choice*, Vol. 36, pp. 235-252.

Call, S.T. and Holahan, W.L., (1983), *Microeconomics*, second edition, Belmount, Calif: Wadsworth.

Carter, M., (1979), "Issues in the Hidden Economy - A Survey", *The Economic Record*, September, pp. 209-221.

*Catalogue of Government Department/Statutory Board Publications 1986*, (1986), Singapore: Ministry of Communications and Information.

Chadeau, A., (1985), "Measuring Household Activities: Some International Comparisons", *The Review of Income and Wealth*, No. 3, pp. 237-253.

Chadeau, A. and Roy, C., (1986), "Relating Households' Final Consumption to Household Activities: Substitutability or Complementary Between Market and Non-market Production", *The Review of Income and Wealth*, December, No. 4, pp. 387-407.

225

Chase Manhattan Bank, (1972), *What is a Wife Worth?* New York: Mimeographed.

Chiswick, B.R., (1967), "The Economic Value of Time and the Wage Rate: Comment", *Western Economic Journal*, Vol. 5, pp. 294-295.

Chiswick, C.V., (1982), "Communications - The Value of a Housewife's Time", *The Journal of Human Resources*, Vol. 17, 3, pp. 413-425.

Clark, C., (1958), "The Economics of Housework", *Bulletin of the Oxford Institute of Statistics*, May, pp. 205-211.

Cogan, J.E., (1975), "Labour Supply and the Value of the Housewife's Time", Office of Economic Opportunity, the Economic Development Administration and the Rockefeller Foundation, pp. 1-42.

Cymrot, D.J. and Seiver, D.A., (1982), "A Note on Individual Utility Maximization in a Household Context", *Eastern Economic Journal*, Vol. 8, pp. 211-214.

Davis, R., (1963), "Recreation Planning as an Economic Problem", *Natural Resources Journal*, Vol. 13, pp. 239-249.

Deegan Jr., J. and White, K.J., (1976), "An Analysis of Nonpartisan Election Media Expenditure Decision Using Limited Dependent Variable Methods", *Social Science Research*, Vol. 5, pp. 127-135.

DeGrazia, S., (1962), *Of Time Work and Leisure*, New York: The Twentieth Century Fund.

Denison, E.F., (1971), "Welfare Measurement and the GNP", *Survey of Current Business*, January issue, pp. 13-16.

DeSerpa, A.C., (1971), "A Theory of the Economics of Time", *Economic Journal* 75, pp. 828-846.

DeSerpa, A.C. and Huntington, H.G., (1978), "Time Allocation, Time Value and Factor Intensity", *Australian Economic Papers*, June, pp. 124-131.

Dhawn, S.L., *Singapore Income Tax*, Criterion Publishers.

*Economic and Social Statistics Singapore 1960-1982*, (1983), Singapore: Department of Statistics.

*Economic Road Maps*, (1984), "Women: A Progress Report, Nos. 1976-1977, published by The Conference Board of Canada.

*Economic Survey of Singapore 1986*, (1986), Singapore: Ministry of Trade and Industry.

Eisner, R., (1978), "Total Incomes in the United States, 1959 and 1969", *The Review of Income and Wealth*, Vol. 24, pp. 41-70.

Evans, A.W., (1972), "On the Theory of the Valuation and Allocation of Time", *Scottish Journal of Political Economy*, February, pp. 1-17.

Ferber, M.A., (1975), "Notes on Maurice Weinrobe's Household

Production: An Improvement of the Record", *Review of Income and Wealth*, Series 21, No. 2, pp. 251-252.

Ferber, M.A., (1980), "Economics of the Family: Who Maximizes What?", *Family Economics Review*, Fall, pp. 13-16.

Ferber, M.A. and Birnbaum, B.G., (1977), "The 'New Home Economics': Retrospects and Prospects", *Journal of Consumer Research*, Vol. 4, pp. 19-28.

Ferber, M.A. and Birnbaum, B.G., (1980), "Housework: Priceless or Valueless?", *The Review of Income and Wealth*, Vol. 28, pp. 387-399.

Ferber, M.A. and Greene, C.A., (1983), "Housework vs. Marketwork: Some Evidence How the Decision is Made", *The Review of Income and Wealth*, June, pp. 147-159.

Finland, Ministry of Social Affairs and Health, Research Department, 1980-81, *Housework Study*, Official Statistics of Finland, Special Social Studies, Vol. 32, 71, 4 parts, Helsinki.

Firebaugh, F.M., and Deacon, R.E., (1980), "Contribution of Women to Development of the Family and the Economy", in C. Hefferan (ed.), *The Household As Producer: A Look Beyond the Market*, American Home Economics Association, pp. 57-71.

Folbre, N., (1984), "Household Production in the Philippines: A non-neoclassical Approach", *Economic Development and Cultural Change*, Vol. 32, pp. 303-329.

Galbraith, J.K., (1973), *Economics and the Public Purpose*, Boston: Houghton Mifflin.

Gauger, W., (1973), "Household Work: Can We Add It to the GNP?", *Journal of Home Economics*, Vol. 65, pp. 12-15.

General Electric Company, (1952), *The Homemaking Habits of the Working Wife*, New York.

Gerner, J.L. and Zick, C.D., (1983), "Time Allocation Decisions in Two-Parent Families", *Home Economics Research Journal*, Vol. 12, No. 2, pp. 145-158.

Goldschmidt-Clermont, L., (1983), "Does Housework Pay? A Product-Related Microeconomic Approach", *Signs: Journal of Women in Culture and Society*, Vol. 9, No. 1, pp. 108-119.

Goldschmidt-Clermont, L., (1983), "Output-Related Evaluations of Unpaid Household Work: A Challenge for Time Use Studies", *Home Economics Research Journal*, Vol. 12, No. 2, pp. 127-132.

Goldschmidt-Clermont, L., (1985), *Unpaid Work in the Household - A Review of Economic Evaluation Methods*, Geneva: International Labour Office.

Graham, J.W. and Green, C.A., (1983), "Estimating The Parameters of a Household Production Function with Joint Products", *The Review of Economics and Statistics*, July, pp. 277-282.

Gramm, W. L., (1974), "The Demand for the Wife's Non-Market Time", *Southern Economic Journal*, Vol. 41, pp. 124-133.

Gronau, R., (1973), "The Effect of Children on the Housewife's Value of Time", *Journal of Political Economy*, Vol. 81, No. 2, Part 2, S169-S199.

Gronau, R., (1973), "The Infrafamily Allocation of the Housewives' Time", *American Economic Review*, Vol. 63, No. 4, pp. 534-651.

Gronau, R., (1980), "Home Production - A Forgotten Industry", *The Review of Economics and Statistics*, Vol. 62, pp. 408-416.

Gunderson, M., (1977), "Logit Estimates of Labour Participation Based on Census Cross-Tabulations", *Canadian Journal of Economics*, Vol. 10, No. 3.

Gunderson, M., (1980), "Probit and Logit Estimates of Labour Force Participation", *Industrial Relations*, Vol. 9, No. 2, pp. 216-220.

Hall, F.T. and Schroeder, M.P., (1970), "Effects of Family and Housing Characteristics on Time Spent on Household Tasks", *Journal of Home Economics*, Vol. 62, No. 1, pp. 23-29.

Hartog, J. and Theevwes, J. (1986), "Participation and Hours of Work - Two Stages in the Life-Cycle of Married Women", *European Economic Review*, Vol. 30, pp. 833-857.

Hawrylyshyn, O., (1974), *A Review of Recent Proposals for Modifying and Extending the Measure of GNP*, Ottawa: Statistics Canada.

Hawrylyshyn, O., (1976), "The Value of Household Services: A Survey of Empirical Estimates", *The Review of Income and Wealth*, Vol. 22, pp. 101-131.

Hawrylyshyn, O., (1977), "Towards a Definition of Non-market Activities", *The Review of Income and Wealth*, Vol. 23, pp. 79-97.

Hawrylyshyn, O., (1978), "Estimating the Value of Household Work in Canada 1971", *Statistics Canada*, Ministry of Industry, Trade and Commerce, Canada, pp. 1-57.

Hawrylyshyn, O. and Woroby, T.; (1982), "The Value of Household Work: Application of the Statistic to Economic and Social Issues"; Paper presented at the Eastern Economic Association Meetings; Washington, D.C.; pp. 1-24.

Heckman, J.J., (1974), "Shadow Prices, Market Wages, and Labour Supply", *Econometrica*, Vol. 42, 4, pp. 679-694.

Hefferan, C., (1982), "What is a Homemaker's Job Worth? Too Many Answers", *Journal of Home Economics*, Vol. 3, pp. 30-33.

228

Hefferan, C., (1986), ed., *The Household as Producer - A Look Beyond the Market*, Proceedings of a workshop sponsored by the Family Economics-Home Management Section of the American Home Economics Association.

Hershlag, Z.Y., (1960), "The Case of Unpaid Domestic Service", *Economia Internazionale*, Vol. 13, pp. 25-48.

Hicks, J.R., (1940), "The Valuation of the Social Income", *Economica*, February, pp. 105-124.

Hicks, J.R., (1942), *The Social Framework*, Oxford: Claredon Press.

Hill, P.T., "Do-It-Yourself and GDP", *The Review of Income and Wealth*, Vol. 25, pp. 31-39.

Ho, T.J., (1976), "Time Allocation, Home Production and Labour Force Participation of Married Women: An Exploratory Survey", Discussion Paper No. 76-8, Institute of Economic Development and Research, School of Economics, University of the Philippines, pp. 1-39.

Hunt, J.C.; DeLorme Jr, C.D.; and Hill, R.C.; (1981), "Taxation and the Wife's Use of Time", *Industrial and Labour Relations Review*, Vol. 34, No. 3.

Jaszi, G., (1973), "Comment on 'A Framework for the Measurement of Economic and Social Performance' - F. Thomas Juster", *The Measurement of Economic and Social Performance*, Studies in Income and Wealth, Vol. 38, edited by Milton Moss, New York: Columbia University Press for the National Bureau of Economic Research.

Joll, C.; McKenna, C; and Shorey, J.; (1983), *Developments in Labour Market Analysis*, Allen and Unwin.

Johnston, J., (1972), *Econometric Methods*, second edition, New York: McGraw-Hill.

Johnston, P.J. and Firebaugh, F.M., (1985), "A Typology of Household Work Performance by Employment Demands", *Journal of Family Issues*, Vol. 6, No. 1, pp. 83-105.

Joseph, G., (1983), *Women At Work - The British Experience*, Philip Allan.

Kahne, H. and Kohen, A.I., (1975), "Economic Perspectives on the Roles of Women in the American Economy", *Journal of Economic Literature*, Vol. 13, pp. 1249-1283.

Kendrick, J.W., (1979), "Expanding Imputed Values in the National Income and Product Accounts", *The Review of Income and Wealth*, Vol. 25, pp. 349-363.

Kennedy, P., (1979), *A Guide to Econometrics*, The MIT Press.

Khoo, C.K., (1981), *Geographic Distribution*, Census of Population, Release No. 5, Singapore: Department of Statistics.

Kilpi , E., (1981), "The Concept of Unpaid Housework and the Determination of Its Value", in Finland, Ministry of Social Affairs and Health, Research Department: *Housework Study*, Part I.

King, W.I. and Epstein, L., (1930), *The National Income and Its Purchasing Power*, NBER Publication No. 15.

Kneeland, H., (1929), "Woman's Economic Contribution in the Home", in *Annuals of the American Academy of Political and Social Sciences*, Vol. 163, pp. 33-40.

Kome, P., (1982), "What Price Dusting? Just How Much is Housework Worth, and Should We Pay for It?" *The Financial Post Magazine*, pp. 55 and 58.

Komesar, N.K., (1974), "Toward a General Theory of Personal Injury Loss", *The Journal of Legal Studies*, Vol. 3, pp. 457-486.

Kuznets, S., (1941), *National Income and Its Composition, 1919-1938*, 2 Volumes, New York: National Bureau of Economic Research.

Lancaster, K., (1966), "A New Approach to Consumer Theory", *Journal of Political Economy*, Vol. 74, pp. 132-157.

Lancaster, K., (1975), "The Theory of Household Behaviour: Some Foundations", *Annuals of Economic and Social Measurement*, 4/1, pp. 5-21.

Leibowitz, A., (1974), "Education and Home Production", *American Economic Review*, Vol. 64, No. 2, pp. 243-250.

Lim, L., (1981), "Women in the Singapore Economy", *ERC Occasional Paper Series*, Vol. 4.

Lindahl, E.; Dahlgren, E. and Kock, K., (1937), *National Income of Sweden, 1861-1930 (in two parts). Vol. III of: Wages, Cost of Living and National Income in Sweden, 1860-1930*, by the staff of the Institute for Social Sciences, University of Stockholm, Stockholm Economic Studies, No. 5a and 5b, Stockholm, Norstedt and S ner.

Mahoney, T.A., (1961), "Influences on Labour Force Participation of Married Women", in Foote, N. (ed.), *Household Decision Making*, New York University Press.

Manning, S.L., (1968), *Time Use in Household Tasks by Indiana Families*, Purdue University Agricultural Experiment Station, Research Bulletin No. 837, Lafayette, Ind.

*Manual of Instructions for Enumerators*, (1980), Census of Population, Singapore: Department of Statistics.

Marshall, A., (1930), *Principles of Economics*, London.

Masoner, M., (1979), "The Allocation of Time: An Extension", *Journal of Post Keynesian Economics*, Vol. 1, No. 3, pp. 107-122.

230

Michael, R.T. and Becker, G.S.,"On the New Theory of Consumer Behavior", *Swedish Journal of Economics*, Vol. 75, No. 4, pp. 378-396.

Michael, R.T., (1985), "Consequences of the Rise in Female Labour Force Participation Rates: Questions and Probes", *Journal of Labour Economics*, Vol. 3, No. 1, pp. S117-S146.

Mincer, J., (1962), "Labour Force Participation of Married Women: A study of Labour Supply", *National Bureau of Economic Research: Aspects of Labour Economics*, Vol. 64, pp. 63-105.

Mitchel, W.; King, W.I.; Macaulay, F.R.; Knauth, O.W.; (1921), *Income in the United States: Its Amount and Distribution, 1909-1919*, National Bureau of Economic Research, publication no. 1, Vol. I, summary, New York: Harcourt, Brace and Co.

Moser, C.A. and Kalton, G., (1971), *Survey Methods in Social Investigation*, second edition, The English Language Book Society and Heinemann Educational Books.

Moss, M. (ed.), (1973), *The Measurement of Economic and Social Performance*, Studies in income and wealth, Vol. 38, by the Conference on Research in Income and Wealth, National Bureau of Economic Research, New York: Columbia University Press.

Muellbauer, J., (1974), "Household Production Theory, and the Hedonic Technique", *The American Economic Review*, Vol. 64, pp. 977-944.

Murphy, M., (1976), "The Value of Time Spent in Home Production", *American Journal of Economics and Sociology*, Vol. 35, pp. 191-197.

Murphy, M., (1978), "The Value of Nonmarket Household Production: Opportunity Cost Versus Market Cost Estimates", *The Review of Income and Wealth*, Series 24, No. 3, pp. 243-255.

Murphy, M., (1980), "Are Household Services Overvalued?" *American Journal of Economics and Sociology*, Vol. 39, pp. 413-415.

Murphy, M., (1982), "Comparative Estimates of the Value of Household Work in the United States for 1976", *The Review of Income and Wealth*, March, pp. 29-43.

Murray, G., (1986), "Singapore Domestic Squabble", *PHP Intersect Where Japan Meets Asia and the World*, Vol. 2, No. 12, pp. 36-37.

Navera, E.R., (1978), "The Allocation of Household Time Associated With Children in Rural Households in Laguna, Philippines", *The Philippine Economic Journal 36*, Vol. 17, Nos. 1 and 2, pp. 203-223.

Nerlove, M., (1974), "Household and Economy: Toward a New Theory of Population and Economic Growth", *Journal of Political Economy*, Vol. 82, No. 2, Part II, S200-S233.

NNW Measurement Committee, Economic Council of Japan, (1974),

*Measuring Net National Welfare of Japan*, Printing Bureau, Ministry of Finance.

Nordhaus, W. and Tobin, J., (1972), "Is Growth Obsolete?" *Economic Growth*, Fiftieth Anniversary Colloquium V., New York: National Bureau of Economic Research.

Norton, M.J.T. and Wall, V.J., (1984), "Evaluation of Research in Home Economics: Background and New Approaches", *Home Economics Research Journal*, Vol. 12, No. 4, pp. 435-449.

Olson, J.T., (1981), "The Impact of Housework on Child Care in the Home", *Family Relations*, Vol. 30, No. 1, pp. 75-81.

Ontario Status of Women Council, (1977), *About Faces: Towards a Positive Image of Housewives*, Canada.

*Ottawa Journal*, (1966), "The Value of a Housewife", January 25, pp. 7.

Pang, E.F., (1973), "A Note on Labour Underutilization in Singapore", *Malayan Economic Review*, Vol. 18, No. 1, pp. 15-23.

Paolucci, B., (1977), "Invisible Family Production: The Development of Human Resources", Paper presented at National Science Foundation.

Peskin, J. and Peskin, H.M., (1978), "The Valuation of Nonmarket Activities in Income Accounting", *The Review of Income and Wealth*, Series 24, No. 1, pp. 71-91.

Peskin, J., (1981), "Measuring Household Production for the GNP", Conference Paper, 1982 Agricultural Outlook Conference, Session No. 10, Washington, D.C., pp. 1-14; *Family Economics Review*, 1982(3), pp. 16-25.

Pigou, A.C., (1932), *Economics of Welfare*, 4th ed., London: Macmillan and Company Ltd.

Pindyck, R.S. and Rubinfield, D.L., (1981), *Econometric Models and Economic Forecasts*, second edition, New York: McGraw-Hill.

Pollak, R.A. and Watcher, M.L., (1975), "The Relevance of the Household Production Function and Its Implications for the Allocation of Time", *Journal of Political Economy*, Vol. 83, No, 2, pp. 255-277.

Proulx, M., (1978), *Women and Work: Five Million Women - A Study of the Canadian Housewife*, Advisory Council on the Status of Women, Canada.

Posner, R.A., (1977), *Economic Analysis of Law*, second edition, Boston: Little, Brown and Company.

Pottick, F., (1978), "Tort Damages for the Injured Homemarker: Opportunity Cost or Replacement Cost?" *University of Colorado Law Review*, Vol. 50, pp. 59-74.

Pyun, C.S., (1969), "The Monetary Value of a Housewife", *American*

*Journal of Economics and Sociology*, July, pp. 271-284.

Quah, Euston, (1982), "Measuring the Amount and Economic Value of Time Devoted to Household Production: An Exploratory Study", Mimeo, Simon Fraser University.

Quah, Euston, (1985), "Household Production and the Measurement of Economic Welfare", *The Indian Journal of Economics*, Oct. Issue, Vol. 66, No. 261, pp. 243-258.

Quah, Euston, (1986a), "Persistent Problems in Measuring Household Production", *The American Journal of Economics and Sociology*, April Issue, Vol. 45, No. 2, pp. 235-246.

Quah, Euston, (1986b), "A Diagrammatic Exposition of Household Production, Market Substitutes and the Terms of Trade", *Singapore Economic Review*, April Issue, Vol. 32, No. 2, pp. 33-39.

Quah, Euston, (1987), "Valuing Family Household Production: A Contingent Evaluation Approach", *Applied Economics*, Vol. 19, July Issue, pp. 875-889.

Quah, Euston, "Exact Compensation in Law and Economics: The Theory of Welfare Loss in Household Production", *Osgoode Hall Law Journal*, (to appear, 1987).

Quah, Euston, (1986c), "What is the Value of a Wife?" *Singapore Business*, February Issue, pp. 15-20.

Quah, Euston, (1986d), "The Economics of Marriage", *Singapore Business*, March Issue, pp. 58-60.

Quizon, E.K., (1978), "Time Allocation and Home Production in Rural Philippine Households", *The Philippine Economic Journal 36*, Vol. 17, Nos. 1 and 2, pp. 185-202.

Reid, M.G., (1934), *Economics of Household Production*, New York: John Wiley and Son.

Reid, M.G., (1947), "The Economic Contribution of Homemakers", *The Annuals of the American of Political and Social Science*, CCLI, pp. 29-33.

*Report on the Household Expenditure Survey 1982-1983*, (1985), Singapore: Department of Statistics.

Robinson, J.P.; Converse, P.E. and Szalai, A., (1972), "Everyday Life in Twelve Countries", in *The Use of Time*, by Szalai, A. (eds.), The Hague, Netherlands' Mouton, pp. 114-143.

Robinson, J. (1977), *How Americans Use Time: A Social-psychological Analysis of Behavior*, New York: Praeger.

Rosen, H.S., (1974), "The Monetary Value of a Housewife: A Replacement Cost Approach", *American Journal of Economics and*

*Sociology* (New York), Vol. 33, No. 1, pp. 65-73.

Ross, R.T., (1982), "General Report of the 1980 Survey of Work Patterns of Married Women in the Sydney Metropolitan Region", Working Papers in Economics No. 62, Department of Economics, The University of Sydney, pp. 1-33.

Ross, R.T., (1985), "Analysis of the 1980 Sydney Survey of Work Patterns of Married Women: Further Results", Working Papers in Economics No. 79, Department of Economics, The University of Sydney, pp. 1-29.

Ruggles, N.D. and Ruggles, R., (1970), *The Design of Economic Accounts*, New York: National Bureau of Economic Research.

Samuelson, P.A., (1956), "Social Indifference Curves", *Quarterly Journal of Economics*, Vol 70, pp. 9 and 21.

Sanik, M.M and Stafford, K.; (1983), "Product-Accounting Approach to Valuing Food Production", *Home Economics Research Journal*, Vol. 12, No. 2, pp. 217-219.

Saw, S.H., (1981), *A Guide to the Economic and Social Statistics of Singapore*, Singapore: Singapore University Press.

Saw, S.H., (1984), *The Labour Force of Singapore*, Census Monograph No. 3, Singapore: Department of Statistics.

Scoggins, J.F., (1987), "Welfare Evaluation and Household Production with Non-Constant Returns to Scale", *Southern Economic Journal*, Vol. 53, No. 3, pp. 643-649.

Scott, A.C., (1972), "The Value of Housework: For Love or Money", *Ms Magazine*.

Scholl, K.K. and Tippett, K.S., eds. (1982), *Family Economic Review Special Issue: Household Production*; Washington, D.C.; United States Department of Agriculture.

Schulze, W.D., et al. (1981), "Valuing Environmental Commodities: Some Recent Experiments", *Land Economics*, Vol. 57, pp. 151-172.

Sen, A., (1984), "Economics and the Family", *Asian Development Review*, Vol. 3, pp. 14-26.

Shamseddine, A.H., (1968), "GNP Imputations of the Value of Housewives' Services", *Economic and Business Bulletin*, Vol. 20, No. 4, pp. 52-61.

Sharir, S., (1975), "The Income-Leisure Model: A Diagrammatic Extension", *Economic Record*, Vol. 51, pp. 93-97.

Sinden, J.A. and Worrell, A.C., (1979), *Unpriced Values - Decisions Without Market Prices*, New York: John Wiley and Sons.

*Singapore Yearbook of Labour Statistics*, (1986), Research and Statistics Department, Singapore: Ministry of Labour.

*Singapore National Accounts 1960-1973*, (1975), Singapore: Department of Statistics.

Sirageldin, I.A., (1969), *Non-market Components of National Income*, Ann Arbor: Survey Research Centre, Institute for Social Research, The University of Michigan.

Stone, R., (1986), "Nobel Memorial Lecture 1984 - The Account of Society", *Journal of Applied Econometrics*, Vol. 1, pp. 5-28.

Stone, R., (1986), "Social Accounting: The State of Play", *Scandinavian Journal of Economics*, Vol. 88, 3, pp. 453-472.

Studenski, P., (1958), *The Income of Nations*, Washington Square: New York University Press.

Suviranta, A. and Heinonen, M., (1980), "The Value of Unsalaried Home Care of Children Under the age of Seven in 1979", in Finland, Ministry of Social Affairs and Health, Research Department: *Housework Study*, Part III.

Suviranta, A. and Mynttinen, A., (1981), "The Value of Unpaid House-Cleaning in 1980", in Finland, Ministry of Social Affairs and Health, Research Department: *Housework Study*, Part IV.

Suviranta, A., (1982), "Housework Study Part VIII - Unpaid Housework: Time Use and Value", Official Statistics of Finland Special Social Studies, Ministry of Social Affairs and Health, Helsinki, pp. 1-46.

Szalai, A.; Ferfe, S.; Goguel; Patrouchev, V.; Raymond, H.; Scheuch, E. and Schneider, A., (1966), "Multinational Comparative Social Research", *American Behaviorial Scientist* 10, Appendix, A/1-A/60.

Szalai, A., (1972), ed., *The Use of Time*, The Hague: Mouton Press.

Szalai, A., (1975), "Women in the Light of Contemporary Time-Budget Research", *Women's Time*, Vol. 10, pp. 385-399.

Szalai, A., (1986), "Trends on Comparative Time-Budget Research", *The American Behavioral Scientist*, Vol. 10, pp. 3-8.

*The Singapore Economy: New Directions*, (1986), Singapore: Ministry of Industry and Trade.

Vanek, J., (1974), "Time Spent in Housework", *Scientific American*, Vol. 231, pp. 116-120.

Vickery, C., (1978), "The Time-poor: A New Look at Poverty" *Journal of Human Resources* (Madison, Wisc.), Vol. 12, pp. 27-48.

Wales, T.J. and Woodland, A.D., (1974), "Estimation of the Allocation of Time for Work, Leisure and Housework", Discussion Paper 74-19, Department of Economics, University of British Columbia, pp. 1-25; and also *Econometrica*, (1977), Vol 45, No. 1, pp. 115-132.

Walker, K.E., (1969), "Homemaking Still Takes Time", *Journal of Home*

*Economics*, Vol. 61, No. 8, pp. 621-624.

Walker, K.E. and Gauger, W.H., (1973), "Time and Its Dollar Value in Household Work", *Family Economics Review*, Fall, pp. 8-13.

Walker, K.E. and Woods, M.E., (1976), *Time Use: A Measure of Household Production of Family Goods and Services*, Washington, D.C.; Centre for the Family of the American Home Economics Association.

Walker, K.E., (1980), "Time Measurement and the Value of Non-market Household Production", in C. Hefferan (ed.), *The Household as Producer: A Look Beyond the Market*, pp. 119-138.

Wannacott, T.H. and Wannacott, R.J., (1977), *Introductory Statistics*, third edition, John Wiley and Sons.

Weinrobe, M., (1974), "Household Production and National Production: An Improvement of the Record", *The Review of Income and Wealth*, Vol. 20, pp. 89-102.

Windschuttle, E., (1972), "Should Government Pay a Mother's Wage?" *Refractory Girl*, pp. 5.

Winegarden, C.R., (1984), "Women's Fertility, Market Work and Marital Status: A Test of the New Household Economics With International Data", *Economica*, Vol. 51, pp. 447-456.

Wong, A.K., (1975), *Women in Modern Singapore*, Singapore: University Education Press.

Wong, A.K., (1980), *Economic Development and Women's Place - Women in Singapore*, Calvert's North Star Press.

Yeh, S.H.K., (1984), *Households and Housing*, Census Monograph No. 4, Singapore: Department of Statistics.

Zick, C.D. and Bryant, W.K., (1983), "Alternative Strategies for Pricing Home Work Time", *Home Economics Research Journal*, Vol. 12, pp. 133-144.